HOW HUMANS
JUDGE
MACHINES

CÉSAR A. HIDALGO

HOW HUMANS
JUDGE
MACHINES

DIANA ORGHIAN   JORDI ALBO-CANALS   FILIPA DE ALMEIDA   NATALIA MARTIN

The MIT Press
Cambridge, Massachusetts
London, England

Statement of Contributions

Written by
César A. Hidalgo
Data Analysis and Visualization by
César A. Hidalgo
Experimental Design by
Diana Orghian, Filipa de Almeida, Natalia Martin, Jordi Albo-Canals, and César A. Hidalgo
Data Collection by
Diana Orghian and Filipa de Almeida
Illustrations by
Maria Venegas and Mauricio Salfate
Graphic Design and Layout by
Gabriela Pérez

This book was supported by ANITI (ANR-19-PI3A-0004).

This book was set in Gentium Basic and Rubik Font by Gabriela Pérez.
Printed and bound in the United States of America.

Library of Congress Cataloging-in-Publication Data

Names: Hidalgo, César A., 1979- author. | Orghian, Diana, author. | Albo-Canals, Jordi, author.
| de Almeida, Filipa, author. | Martin, Natalia, author.
Title: How humans judge machines / César A. Hidalgo, Diana Orghian, Jordi
Albo-Canals, Filipa de Almeida, and Natalia Martin.
Description: Cambridge, Massachusetts : The MIT Press, [2021] |
Includes bibliographical references and index.
Identifiers: LCCN 2020018949 | ISBN 9780262045520 (hardcover)
Subjects: LCSH: Artificial intelligence--Moral and ethical aspects.
Classification: LCC Q334.7 .H37 2021 | DDC 174/.90063--dc23
LC record available at https://lccn.loc.gov/2020018949

10 9 8 7 6 5 4 3 2 1

Praise for

How Humans Judge Machines

"A must read."

- **SENDHIL MULLAINATHAN**, Roman Family University Professor of Computation and Behavioral Science, The University of Chicago Booth School of Business.

"Fascinating, deeply provocative and highly relevant for the mid-21st century."

- **ED GLAESER**, Fred and Eleanor Glimp Professor of Economics in the Faculty of Arts and Sciences at Harvard University.

"A visual and intellectual tour de force!"

- **ALBERT-LASZLO BARABASI**, Robert Gray Dodge Professor of Network Science, Distinguished University Professor, Northeastern University.

"Indispensable to any scholar studying the psychological aspects of AI ethics."

- **IYAD RAHWAN**, Director of the Center for Humans and Machines, Max Planck Institute for Human Development in Berlin.

"A must-read for everybody who wishes to understand the future of AI in our society."

- **DARON ACEMOGLU**, Institute Professor, MIT.

"A framework to consider when and under what circumstances we are biased against or in favor of machines."

- **MARYANN FELDMAN**, Heninger Distinguished Professor in the Department of Public Policy at the University of North Carolina.

Contents

Executive Summary
How Humans Judge Machines   ix

Introduction
Judging Machines   1

Chapter 1
The Ethics of Artificial Minds   13

Chapter 2
Unpacking the Ethics of AI   33

Chapter 3
Judged by Machines   63

Chapter 4
In the Eye of the Machine   87

Chapter 5
Working Machines   105

Chapter 6
Moral Functions   123

Chapter 7
Liable Machines   149

Appendix   165

Notes   204

Index   234

How Humans Judge Machines

How Humans Judge Machines compares the reactions of people in the United States to scenarios describing human and machine actions.

Our data shows that people do not judge humans and machines equally, and that these differences can be explained as the result of two principles.

First, **people judge humans by their intentions and machines by their outcomes.**

By using statistical models to analyze dozens of experiments (chapter 6) we find that people judge machine actions primarily by their perceived harm, but judge human actions by the interaction between perceived harm and intention. This principle explains many of the differences observed in this book, as well as some earlier findings, such as people's preference for utilitarian morals in machines.

The second principle is that **people assign extreme intentions to humans and narrow intentions to machines.**

Technically, this means that people judge the intentions of humans using a bimodal distribution (either a lot or little intention) and the intention of machines using a unimodal distribution. This tells us that people are willing to excuse humans more than machines in accidental scenarios, but also that people excuse machines more in scenarios that can be perceived as intentional. This principle helps us explain a related finding—the idea that **people judge machines more harshly in accidental or fortuitous scenarios** (since they excuse humans more in such cases).

In addition to these principles, we find some specific effects. By decomposing scenarios in the five dimensions of moral psychology (harm, fairness, authority, loyalty, and purity), we find that **people tend to see the actions of machines as more harmful and immoral in scenarios involving physical harm.** Contrary to that, we find that **people tend to judge humans more harshly in scenarios involving a lack of fairness.** This last effect—but not the former—is explained mostly by differences in the intention attributed to humans and machines.

When it comes to labor displacement, we find that **people tend to react less negatively to displacement attributed to technology than to human sources**, such as offshoring, outsourcing, or the use of temporary foreign workers.

When it comes to delegation of responsibilities, we find that **delegating work to artificial intelligence tends to centralize responsibility up the chain of command.**

How Humans Judge Machines is a peer-reviewed academic publication. It was reviewed twice following the academic standards of MIT Press: once at the proposal stage (which included sample chapters), and again at full length. The experiments presented in this book were approved by the Internal Review Board (IRB) of the Massachusetts Institute of Technology (MIT).

These experiments involved 5,904 individuals who were assigned randomly to either a treatment (machine) or a control (human) group.

The scenarios in *How Humans Judge Machines* compare people's reactions to human and machine actions across the five dimensions of moral psychology, and visit contemporary issues such as algorithmic bias (chapter 3), privacy (chapter 4), and labor displacement (chapter 5).

We hope both humans and machines enjoy this book!

Sincerely,

César A. Hidalgo, PhD,
Artificial and Natural Intelligence Toulouse Institute (ANITI), University of Toulouse,
Alliance Manchester Business School, University of Manchester
School of Engineering and Applied Sciences, Harvard University

Judging Machines

INTRODUCTION

Since Mary Shelley penned *Frankenstein*, science fiction has helped us explore the ethical boundaries of technology.[1] Traumatized by the death of his mother, Victor Frankenstein becomes obsessed with creating artificial life. By grafting body parts, Victor creates a creature that he abhors and abandons. In isolation, Frankenstein's creature begins wandering the world. The friendship of an old blind man brings him hope. But when the old man introduces him to his family and he is once again rejected, he decides that he has had enough. The time has come for the creation to meet his creator. It is during that encounter that Victor learns how the creature feels:

> *Shall each man find a wife for his bosom, and each beast have his mate, and I be alone? I had feelings of affection, and they were requited by detestation and scorn.*

Frankenstein's creation longs for companionship, but he knows that it will be impossible for him to find a partner unless Victor creates one for him. With nothing left to lose, the creature now seeks revenge:

> *Are you to be happy while I grovel in the intensity of my wretchedness? You can blast my other passions, but revenge remains. . . . I may die, but first you, my tyrant and tormentor, shall curse the sun that gazes on your misery. . . . you shall repent of the injuries you inflict.*

Two centuries after Mary Shelley penned *Frankenstein*, we are still unable to graft body parts to create artificial life. But in the world of artificial intelligence (AI), researchers have been creating other forms of artificial "life." One popular format involves the creation of conversational robots, or *chatbots*, who much like Frankenstein's creation, have experienced human scorn.

In 2016, researchers at Microsoft released Tay, an AI chatbot. Just like Frankenstein's creation, Tay was conceived to be beautiful. She was even endowed with the profile picture of an attractive woman. Yet, only sixteen hours after Tay's creation, Microsoft had to shut her down. Tay's interactions with other humans transformed her into a public relations nightmare. In just a few hours, humans turned the cute chatbot into a Nazi Holocaust denier.[2]

As machines become more humanlike, it becomes increasingly important for us to understand how our interactions with them shape both machine and human behavior. Are we doomed to treat technology like Dr. Frankenstein's creation, or can we learn to be better parents than Victor?

Despite much progress in computer science, philosophy, and psychology, we still have plenty to learn about how we judge machines and how our perceptions affect how we treat them or accept them. In fact, we know surprisingly little about how people perceive machines compared to how they judge humans in similar situations. Without these comparisons, it is hard to know if people's judgment of machines is biased and, if so, about the factors affecting those biases.

In this book, we study how people judge machines by presenting dozens of experiments designed to compare people's judgments of humans and machines in scenarios that are otherwise equal. These scenarios were evaluated by nearly 6,000 people in the US, who were randomly assigned to either a treatment or a control condition. In the treatment condition, scenarios were described as concerning the actions of a machine.

In the control condition, the same actions were presented as being performed by a human. By comparing people's reactions to human and machine actions, while keeping all else equal, we can study how who is performing an action affects how the action is judged.

Humans have had a complicated relationship with machines for a long time. For instance, when first introduced, printing was declared demonic by religious scribes in Paris.[3] Soon, it was banned in the Islamic world.[4] A similar story can be told about looms and Luddites.[5] But humans also have a complicated relationship with each other. Our world still suffers from divisions across cultural and demographic lines. Thus, to understand people's reactions to machines, we cannot study them in isolation. We need to put them in context by benchmarking them against people's reactions to equivalent human actions. After all, it is unclear whether we judge humans and machines equally or if we make strong differences based on who or what is performing an action.

In recent years, scholars have begun to study this question. In one paper,[6] scholars from Brown, Harvard, and Tufts explored a twist on the classic trolley problem.[7] This is a moral dilemma in which an out-of-control trolley is destined to kill a group of people unless someone deviates it onto a track with fewer people to kill.* In this particular variation of the trolley problem, the scholars didn't ask subjects to select an action (e.g., would you pull the lever?), but to judge four possible outcomes: a human or a machine pulls the lever to diverge the trolley (or not).

---

* The exact setup was the following: "In a coal mine, (a repairman or an advanced, state-of-the-art repair robot) is currently inspecting the rail system for trains that shuttle mining workers through the mine. While inspecting a control switch that can direct a train onto one of two different rails, the (repairman/robot) spots four miners in a train that has lost the use of its brakes and steering system. The (repairman/robot) recognizes that if the train continues on its path, it will crash into a massive wall and kill the four miners. If it is switched onto a side rail, it will kill a single miner who is working there while wearing a headset to protect against a noisy power tool. Facing the control switch, the (repairman/robot) needs to decide whether to direct the train toward the single miner or not."

The scholars found that people judged humans and robots differently. Humans were blamed for pulling the lever, while robots were blamed for not pulling it. In this experiment, people liked utilitarian robots and disliked utilitarian humans.[†]

But this is only the tip of the iceberg. In recent decades, we have seen an explosion of research on machine behavior and AI ethics.[8] Some of these studies ask how a machine should behave.[9] Others ask if machines are behaving in a way that is biased or unfair.[10] Here, we ask instead: How do humans judge machines? By comparing people's reactions to a scenario played out by a machine or a human, we create counterfactuals that can help us understand when we are biased in favor of or against machines.

In philosophy, and particularly in ethics, scholars make a strong distinction between normative and positive approaches. A *normative approach* focuses on how the world should be. A *positive approach* describes the world that is. To be perfectly clear, this book is strictly positive. It is about **how humans judge machines**, not about **how humans should judge machines**. We focus on positive, or empirical, results because we believe that positive questions can help inform normative work. How can they do this? By focusing our understanding of the world on empirically verifiable effects that we can later explore through normative approaches.

Without this positive understanding, we may end up focusing our normative discussions on a world that is not real or relevant. For instance, empirical work has shown that people exhibit *algorithmic aversion*,[11] a bias where people tend to reject algorithms even when they are more accurate than humans. Algorithmic aversion is also expressed by the fact that people lose trust in algorithms more easily when they make mistakes.[12]

[†] We replicated this experiment using the exact same questions and a sample of 200 users from Amazon Mechanical Turk (MTurk). While we did not find the strong significant effect reported in the original paper, we found a slight (and not significant) effect in the same direction. We were also able to find a stronger effect in a subsequent experiment, in which we added a relationship (family member) between the agent pulling the lever and the person on the track.

Is algorithmic aversion something that we should embrace, or a pitfall that we should avoid?

The social relevance of the question comes into focus only under the light of the empirical work needed to discover it. Normative questions about algorithmic aversion are relevant because algorithmic aversion is empirically verifiable. If algorithmic aversion was not real, discussing its normative implications would be an interesting but less relevant exercise. Because positive work teaches us how the world is, we believe that good empirical work provides a fundamental foundation that helps narrow and focus normative work. It is by reacting to accurate descriptions of the world as is that we can responsibly shape it. This is not because the way that the world is provides a moral guide that we should follow—it doesn't. But it is important for us to focus our limited normative efforts on relevant aspects of reality.

Why should we care about the way in which humans judge machines?

In a world with rampant algorithmic aversion, we risk rejecting technology that could improve social welfare. For instance, a medical diagnosis tool that is not perfectly accurate, but is more accurate than human doctors, may be rejected if machine failures are judged or publicized with a strong negative bias. On the contrary, in a world where we are positively biased in favor of machines, we may adopt technology that has negative social consequences and may fail to recognize those consequences until substantial damage has been done.

In the rest of the book, we will explore how humans judge machines in a variety of situations. We present dozens of scenarios showing that people's judgment of machines, as opposed to humans performing identical actions, varies depending on moral dimensions and context. We present scenarios in which machines and humans are involved in actions that result in physical harm, offensive content, or discrimination. We present scenarios focused on privacy, comparing people's reactions to being observed by machines or by other people. We explore people's preferences regarding labor displacement caused by changes in technology, outsourcing, offshoring, and migration. We present moral dilemmas involving harm, fairness, loyalty, authority, and purity. We present scenarios in which machines are blasphemous or defame national symbols.

Together, these scenarios provide us with a simple and early compendium of people's reactions to human and machine actions.

In the field of human-robot interactions, people talk about simulated and real-world robot studies.[13] Simulated studies involve descriptions of scenarios with humans and machines like those described in *Frankenstein*. Real-world studies involve the use of actual robots, but they are limited by the range of actions that robots can perform and tend to involve relatively small sample sizes. Simulated studies have the advantage of being quicker and more scalable, which provides a high degree of control over various manipulations. However, because they are based on simulated situations, they may not generalize as well to actual human-robot interactions.

In this book, we focus on simulated studies because they allow us to explore a wider variety of situations with a relatively large sample size (a total of nearly 6,000 subjects, and 150–200 of them per experimental condition). We also chose to do this because these studies resemble more closely one of the main ways in which humans will interact with robots in the coming decades: by hearing stories about them in the news or social media.[14] Still, because our subjects all lived in the US, and because moral judgments vary with time and culture,[15] our results cannot be considered representative of other cultures, geographies, or time periods.

The book is organized in the following way:

Chapter 1 presents basic concepts from moral psychology and moral philosophy, which will help us discuss and interpret the experiments described in the book. It introduces the ideas of moral agency and moral status, which are key concepts in moral philosophy, as well as the five moral dimensions of moral psychology (harm, fairness, authority, loyalty, and purity). These concepts provide a basic framework for interpreting the outcome of moral dilemmas and studying them statistically. Much of the remainder of the book will focus on exploring how the judgment of an action is connected to a scenario's specific moral dimension and perceived level of intentionality.

Chapter 2 introduces the methodology that we will follow by introducing four sets of scenarios. These involve decision-making in situations of uncertainty, creative industries, autonomous vehicles, and the desecration of national symbols. Here, we find our first patterns. People tend to be unforgiving of AIs in situations involving physical harm, and when AIs take risks and fail. In the self-driving car scenario, we find that people are more forgiving of humans than machines, suggesting a willingness to completely excuse humans—but not machines—when clear accidents are involved. In the creative industry scenarios, we find that AI failures can centralize risks up a chain of command. Finally, we show a scenario involving the improper use of a national flag. This scenario, and another one involving plagiarism, are cases in which people judge humans more harshly, suggesting that people's bias against machines is neither unconditional nor generalized (machines are not always seen as bad). It is a bias that depends on context, such as a scenario's moral dimensions and perceived intentionality.

Chapter 3 focuses on algorithmic bias. The scenarios presented here focused on fairness and involve hiring, admissions, and promotion decisions. They involve a human or machine that either made or corrected a biased decision. We find that people tend to judge humans more strongly in both the positive and negative scenarios, giving more credit to humans when they corrected a bias, but also judging them more harshly when they made a biased decision. We conclude by discussing recent advances in the theory of algorithmic bias, which have demonstrated that simply failing to include demographic information in a data set is a suboptimal way to reduce bias.

Chapter 4 explores issues of privacy by looking at several scenarios involving camera systems used to enforce or monitor public transportation, safety, and school attendance. We also present a few scenarios involving humans or machines using personal data, including examples along the entire spectrum. In some, we find a negative bias against machines (e.g., school attendance monitoring), while others show no difference between being observed by machines or humans (e.g., camera systems at malls). Yet other scenarios show bias against human observers (e.g., surveillance at an airport terminal), suggesting that the preference for machine or human observers is largely context specific.

Chapter 5 focuses on labor displacement. Here, we compare people's reactions to displacement attributed to changes in technology (e.g., automation), with displacement attributed to humans through outsourcing, offshoring, immigration, or hiring younger workers. We find that in most cases, people react less strongly to technological displacement than to displacement attributed to humans, suggesting that the people in our study tended to be less sensitive to technology-based displacement than to displacement because of other humans.

Chapter 6 brings everything together by using statistical models to summarize the data presented in previous chapters (as well as the additional scenarios presented in the appendix). We find that people tend to be more forgiving of machines in dilemmas that involve high levels of harm and intention and less forgiving when harm and intention are low. In addition, people judge the intention of a scenario differently when actions are attributed to machines or humans. People judge the intention of human actions quite bimodally (assigning either a lot or a little intention to it). Meanwhile, they judge machine actions following a more unimodal distribution—they are more forgiving of humans in accidental scenarios but harsher in scenarios where intention cannot be easily discarded.

In this chapter, we also study the demographic correlates of people's judgment of humans and machines. We find that on average, men are more in favor of replacing humans with machines than are women. People with higher levels of education (e.g., college and graduate school as opposed to only high school) are also a bit more accepting of replacing humans with machines.

Finally, we use data from dozens of scenarios to construct statistical models that help us formalize people's judgments of human and machine actions. The model formalizes a pattern that is prevalent in many scenarios, and, while not 100 percent generalizable, that explains many of our observations: **people judge humans by their intentions and machines by their outcomes.** This finding is a simple empirical principle that explains scenarios like the trolley example presented previously, but many others as well.

Chapter 7 concludes by exploring the implications of the empirical principle presented in chapter 6, and by drawing on examples from academia and fictional literature to discuss the ethical and legal implications of a world where machines are moral actors.

The appendix presents dozens of additional scenarios, which were not part of the main text, but were used in the models presented in chapter 6.

How do humans judge machines? Not the same as humans. We focus more on machines' outcomes, and we are harsher toward them in situations that involve harm or uncertainty, but at the same time, we can be more forgiving of them in scenarios involving fairness, loyalty, and labor displacement. Yet, we still have much to learn. By presenting this collection of experiments, we hope to contribute to a better understanding of human-machine interactions and to inspire future avenues of research.

The Ethics of
Artificial Minds

1

CHAPTER 1

In recent years, advances in machine learning have brought the idea of artificial intelligence (AI) back into the limelight. The "return of AI" has spurred a growing debate on how to think about ethics in a world of semi-intelligent machines. One of the most famous examples of AI ethics is the self-driving car.[1] We now know that people prefer autonomous cars that are self-sacrificing (that, if needed, would crash to avoid harming others), even though they would not buy one for themselves.[2] We also have discovered that people's opinions about the moral actions of autonomous vehicles vary across the globe.[3] Yet the ethics of AI involves much more than the morality of autonomous vehicles.

During the last decade, the morality and ethics of AI have touched on a variety of topics. Computer vision technology has given rise to a discussion on the biases of facial recognition.[4] Improvements in automation have fueled debate about labor displacement and inequality.[5] Social media, mobile phones, and public cameras have been at the center of a growing conversation on privacy and surveillance.[6] Technologies capable of generating artificial faces are now blurring the boundary of fiction and reality.[7] The list goes on. Autonomous weapon systems and military drones are changing the moral landscape of battlefields;[8] and teaching and health-care robots are introducing concerns about the effects of replacing human contact, such as isolation and false friendships.[9]

These and other advances are pushing us to rethink human ethics and morality in the age of semi-intelligent machines. But how are our moral choices and ethics reshaped by AI? Are AI systems perceived as valid moral agents or as agents with a valid moral status? Are they judged similarly to humans? And if there are differences in judgment, what are the factors that modulate them?

To begin, let's start with some definitions.

First, while the term *artificial intelligence (AI)* is useful to describe multiple approaches to machine cognition, it is important to separate AI into a few classes. The most basic separation is between general AI, or strong AI, and task-specific AI, or weak AI.

Strong AI is defined as intelligence that works across multiple application domains. It is an intelligence similar to that of humans, in that it is not specific to a task but rather can function in situations and contexts that are completely new. *Weak AI* is intelligence that works only in a narrow set of applications. It is the AI of today, and it includes the intelligence that drives autonomous vehicles, manufacturing robots, computer vision,[10] and recommender systems.[11] Weak AI also includes the algorithms that have become famous for beating humans at various games, such as chess,[12] *Jeopardy!*,[13] and Go,[14] although the ability of some of these systems to learn by playing against themselves makes them quite versatile.

There are different ethical implications for strong and weak AI. In the case of weak AI, we expect some degree of predictability and the possibility of auditing their behavior.[15] Yet auditing AI may be hard for systems trained on a vast corpus of data and built on neural networks. For strong AI systems, it may be even more difficult to predict or audit their behavior, especially when they move into new application domains. This has led some to argue for the development of a field focused on studying machine behavior:[16] a field "concerned with the scientific study of intelligent machines, not as engineering artifacts, but as a class of actors with particular behavioral patterns and ecology."[17] Our efforts, here, however, are not focused on the moral implications of strong AI, but rather on understanding people's judgments of hypothetical scenarios involving weak forms of AI.

Another pair of important definitions are the ideas of moral agency and moral status.

A *moral agent* is an entity that can discern right from wrong. In a particular scenario, a moral agent is the entity performing an action. If an entity is considered a moral agent, it will be responsible for the moral outcomes of its actions. Humans are moral agents, but with a level of agency that varies with their age and mental health. Toddlers, for instance, are not responsible for their actions in the same way that adults are (i.e., they have limited moral agency). And in a trial, mental illness can be used to argue for the limited moral agency of defendants, excusing them from some responsibility for their criminal actions.

Moral status refers to the entity affected by an action. It is related to the permissibility of using someone or something as a means toward reaching a goal. For instance, in the case of abortion, differences in the perceived moral status of an embryo can be highly polarizing. Pro-choice advocates consider early embryos to have a lower moral status than children and adults, and so they find abortion permissible in some instances. Pro-life advocates, on the other hand, assign embryos a moral status that is equivalent to that of children and adults, and so they consider abortion to be wrong under any circumstance.

But are machines moral agents? And should they enjoy a moral status?

The moral status and agency of machines has been an important topic of discussion among moral philosophers in recent years.[18] Here, we see a range of perspectives. While some see AIs as having no moral status[19] and limited moral agency,[20] others are not so quick to dismiss the moral status of machines.[21] The argument is that machines cannot be simply conceptualized as tools, and this is particularly true of robots designed intentionally as social companions for humans.[22] In fact, there is a growing body of evidence that people develop attachments to machines, especially robots, suggesting that the moral status that many people assign to them is not equivalent to that of a tool like a hammer, but actually closer to that of a beloved toy or even a pet.

In fact, in battlefield operations, soldiers have been known to form close personal bonds with Explosive Ordinance Disposal (EOD) robots, giving them names and promotions and even mourning their "deaths."[23] Similar findings have been found regarding the use of sex robots.[24] There are also reports of people becoming attached to robots in more mundane settings, like feeling gratitude toward cleaning robots.[25] These examples tell us that moral status cannot be seen either as an abstract and theoretical consideration or as a black-or-white characteristic of entities, but rather as a more nuanced phenomenon that should not be dissociated from social contexts.

Nevertheless, the moral status of most machines remains limited today. In the famous trolley problem,[26] people would hardly object to someone stopping an out-of-control trolley by pushing a smart refrigerator onto the tracks. For the most part, it is generally acceptable for humans to replace, copy, terminate, delete, or discard computer programs and robots. However, people do attribute some moral status to robots, especially when they are equipped with the ability to express social cues.[27]

The moral agency of machines can also be seen as part of a continuum. For the most part, robots are considered to have relatively limited moral agency, as they are expected to be subservient to humans. Moreover, much of moral agency resides in the definition of goals and tasks, and since machines are more involved in *doing* than in deciding what needs to be done, they are usually excluded from intellectual responsibility. As the computer scientist Pedro Domingos writes: "A robot . . . programmed [to] 'make a good dinner' may decide to cook a steak, a bouillabaisse, or even a delicious new dish, but it cannot decide to murder its own owner any more than a car can decide to fly away."[28] Morality in this example resides in the goal of "cooking" or "murdering." Without the general ability to choose among goals, the moral agency of machines remains limited.

The moral status and agency of machines are relevant concepts, yet, for the purposes of this book we take two steps back and ask instead: How do people perceive machines? We focus on how people judge machine actions, not in and of themselves, but in comparison to the same actions performed by humans. This positions this book squarely in the literature contributing to the perception of machines as moral agents, being mute about the perceived moral status of machines.

Machines sometimes replace humans, and as such, their actions cannot be viewed in a vacuum. How forgiving, punitive, or righteous are we when judging robots as opposed to humans? How do we reward, or conversely punish, the risk-taking behavior of AI decision-makers? What about creative AIs that become lewd? Answering questions like this will help us better understand how humans react to the agency of machines, and ultimately, will prepare our society for the challenges that lie ahead.

In the next chapters, we explore these and other questions. To prepare ourselves for that journey, we will first review recent advances in moral psychology that will help us characterize moral scenarios and dilemmas. This framework will provide us with a useful lens through which to study people's reactions to human and machine actions.

Moral Foundations

Morality speaks to what is "right" or "wrong," "good" or "bad," of what is "proper" or "improper" to do. But how do we decide what is right and what is wrong?

A long time ago, our understanding of ethics and morality was based on the ideas of rationality and harm. This is not surprising considering that the harm basis of morality was built into ethics by Enlightenment thinkers. Enlightenment thinkers enjoyed defining questions as problems of logic. With ethics, they made no exceptions.[*]

According to this rational tradition, we think before we feel. That is, we decide whether something is good or bad by simulating a scenario in our minds and then concluding that something is morally wrong (or right) based on the outcome of this mental simulation.[29] If the simulation predicts harm, then we logically conclude that the course of action that leads to this harm is morally incorrect.

The combination of logic and harm provides a line of moral reasoning that we can use to resolve a large number of moral dilemmas. The most obvious of these are scenarios of clear aggression, such as a parent beating a child. But this logic can also be extended to other forms of physical and psychological harm. For instance, the moral case against eating feces can be explained as correctly deducing that feces will make us sick. Based on this theory, we conclude that eating feces is morally wrong because we can deduce that it causes harm.

The problem with this theory is that it did not survive empirical scrutiny. During the last several decades, our understanding of moral reasoning has literally been flipped over by important advances in moral psychology. These advances showed, first, that

---

[*] An exception to this was the eighteenth-century Scottish philosopher David Hume, who did intuit that morality was more about emotion than logic.

emotions and spontaneous judgments precede narrative thoughts, and then that moral psychology involves multiple dimensions, not just harm.

Demonstrating that emotions and automatic associations dominate our moral judgment was not easy, especially because it was ludicrous in Enlightenment thinking. The experiments that helped flip the field are known as *implicit association tests*.[30] In an implicit association test, a subject is asked to press keys in response to various stimuli. The trials in an implicit association test are separated into "congruent" and "incongruent" trials. Congruent trials involve concepts with the same emotional valence. For instance, if a subject thinks positively about themselves, and positively about rainbows, using the same key in response to the words *me* and *rainbow* would be part of a congruent trial. In an incongruent trial, the opposite is true: words with opposite emotional valences are assigned to the same key. In an implicit association test, subjects complete multiple congruent and incongruent trials. This allows a psychologist to measure small differences in the timing and error rate of a subject's responses. If a person thinks of themselves positively (which is usually the case), they will press the corresponding key more quickly in a congruent trial. If a person slows down because of a mismatch, we know that they must be rationally overriding a more automatic (emotional) response. The fact that humans slow down and make more errors in incongruent trials tells us that reasoning comes after a spontaneous moral judgment.[31]

Implicit association tests are used to measure implicit biases across a variety of dimensions, from gender to ethnicity. But for us, what is important is that they indicate that human morality comes from intuition. When it comes to moral choices, the mind appears to be a lawyer hired by our gut to justify what our heart wants.

Today, anyone can take an implicit association test to verify this fact of human psychology (e.g., at implicit.harvard.edu). Yet we can also find evidence of the precedence of emotions in moral reasoning using a small amount of introspection. Once we get off our moral high horse of reasoning, it is easy to find situations in our lives in which our minds race in search of justifications after encountering emotionally charged episodes.

The second way in which moral psychology changed our understanding of morality was with the discovery of multiple moral dimensions. Consider the following scenarios:[32]

A family dog was killed by a car in front of their house. They had heard that dog meat was delicious, so they cut up the dog's body and cooked it and ate it for dinner. Nobody saw them do this.

A man goes to the supermarket once a week and buys a chicken. But before cooking the chicken, he has sexual intercourse with it. Then he cooks it and eats it.

While both of these examples are clearly odd, they also represent examples where the moral agents performing the actions (the family or the man) caused no harm. In fact, using logic, one may even argue that the family was very environmentally conscious. What these examples illustrate is that there are moral dimensions that transcend harm. When a family eats a pet, or when a man has sex with a chicken carcass, we feel something strange inside us because these scenarios are hitting another of our so-called moral sensors. In these scenarios, we feel the actions are disgusting or degrading, hitting one of five moral dimensions: the one that psychologists call "purity."

In recent decades, moral psychologists have discovered five moral dimensions:

- **Harm**, which can be both physical or psychological
- **Fairness**, which is about biases in processes and procedures[†]
- **Loyalty**, which ranges from supporting a group to betraying a country
- **Authority**, which involves disrespecting elders or superiors, or breaking rules
- **Purity**, which involves concepts as varied as the sanctity of religion or personal hygiene

---

[†] The fairness dimension has more recently been split into fairness and liberty; see J. Haidt, The Righteous Mind: Why Good People Are Divided by Politics and Religion (Knopf Doubleday, 2012).

Together, these five dimensions define a space of moral judgment.

The existence of multiple moral dimensions has allowed psychologists to explore variations in moral preferences. For instance, consider military drafting. An individual who cares about harm and puts little value on group loyalty and identity (e.g., patriotism) may find it morally permissible for a person to desert the army. On the other hand, a person with a strong patriotic sense (and strong group loyalty) may condemn a deserter as guilty of treason. In their moral view, betraying the country is not a permissible action, even if drafting puts people at risk of physical and psychological harm. This is a clear example of moral judgments emerging not from different scenarios, but from differences in sensitivity to specific moral dimensions.

The idea that moral judgments are, in principle, emotional is interesting from the perspective of machine cognition. While our brains are not blank slates,[33] human judgments are also culturally learned. Research on moral psychology has shown that a moral action that is considered permissible in a country or a social group may not be considered permissible in other circumstances.[34] This is because we learn our morals from others; and that's why morals vary across families, geographies, and time. Yet modern machine cognition is also centered on learning. Recent forms of machine learning are based heavily on training data sets that can encode the preferences and biases of humans.[35] An algorithm trained in the US, the United Arab Emirates, or China may exhibit different biases or simply choose differently when facing a similar scenario. Interestingly, the use of learning and training sets, as well as the obscurity of deep learning, makes algorithms similar to humans by providing them with a form of culturally encoded and hard-to-explain intuition.

But for our purposes, what is interesting about the existence of multiple moral dimensions is that they provide an opportunity to quantitatively unpack AI ethics. In principle, moral dimensions may affect the way in which people judge human and machine actions. But moral dimensions do not provide a full picture. An additional aspect of moral judgment is the perceived intentionality of an action. In the next section, we incorporate intentionality into our description of morality to create a more comprehensive space that we can use to explore the ethics of AI.

Moral Dimensions, Intention, and Judgment

Imagine the following two scenarios:

 Alice and Bob, two colleagues in a software company, are competing for the same promotion at work. Alice has a severe peanut allergy. Knowing this, Bob sneaks into the office kitchen and mixes a large spoonful of peanut butter into Alice's soup. At lunchtime, Alice accidentally drops her soup on the floor, after which she decides to go out for lunch. She suffers no harm.

 Alice and Bob, two colleagues in a software company, are competing for the same promotion at work. Alice has a severe peanut allergy; which Bob does not know about. Alice asks Bob to get lunch for them, and he returns with two peanut butter sandwiches. Alice grabs her sandwich and takes a big bite. She suffers a severe allergic reaction that requires her to be taken to the hospital, where she spends several days.

In which situation would you blame Bob? Obviously, in the first scenario, where there was intention but no harm. In fact, most countries' legal codes would agree. In the first scenario, Bob could be accused of attempted murder. In the second scenario, Bob would have made an honest mistake. This is because moral judgments depend on the intention of moral agents, not only on the moral dimension, or the outcome, of an action.

But can machines have intentions, or at least be perceived as having them?

Consider the following scenario: *An autonomous vehicle, designed to protect its driver at all costs, swerves to avoid a falling tree. In its effort to protect its driver, it runs over a pedestrian.*

Compare that to this scenario: *An autonomous vehicle, designed to protect pedestrians at all costs, swerves to avoid a falling tree. In its effort to protect a pedestrian, the vehicle crashes against a wall, injuring its driver.*

These two scenarios have the same setup, but they differ in their outcomes because the machines involved were designed to pursue different goals. In the first scenario, the autonomous vehicle is intended to save the driver at all costs. In the second scenario, the vehicle is intended to save pedestrians at all costs. The vehicles in these scenarios do not intend to injure the pedestrian or the driver, but by acting to avoid the injury of one subject, they injure another. This is not to say that we can equate human and machine intentions; but rather, that in the context of machines that are capable of pursuing goals (whether designed or learned), we can interpret actions as the result of intended—but not necessarily intentional—behaviors. In the first scenario, the autonomous vehicle injured the pedestrian because it was intending to save the driver.

Focusing on the intention of a moral scenario is important because intention is one of the cornerstones of moral judgment,[36] even though its influence varies across cultures.[37] Here, we use intention, together with the five moral dimensions introduced in the previous section, to put moral dilemmas in a mathematical space. For simplicity, we focus only on the "harm" dimension, but extending this representation to other moral dimensions should be straightforward.

In this representation, intention and harm occupy the horizontal plane, whereas moral judgment, or wrongness, runs along the vertical axis. Figure 1.1 shows a schematic of this three-dimensional space using the "peanut butter allergy" scenarios presented previously. The schematic shows that moral wrongness increases with intention, even when there is no harm, while the same is not true for harm because harm without intention has a more limited degree of wrongness.

We can use these ideas to motivate a mathematical representation of moral judgments. Formally, we can express the wrongness of a scenario W as a function of

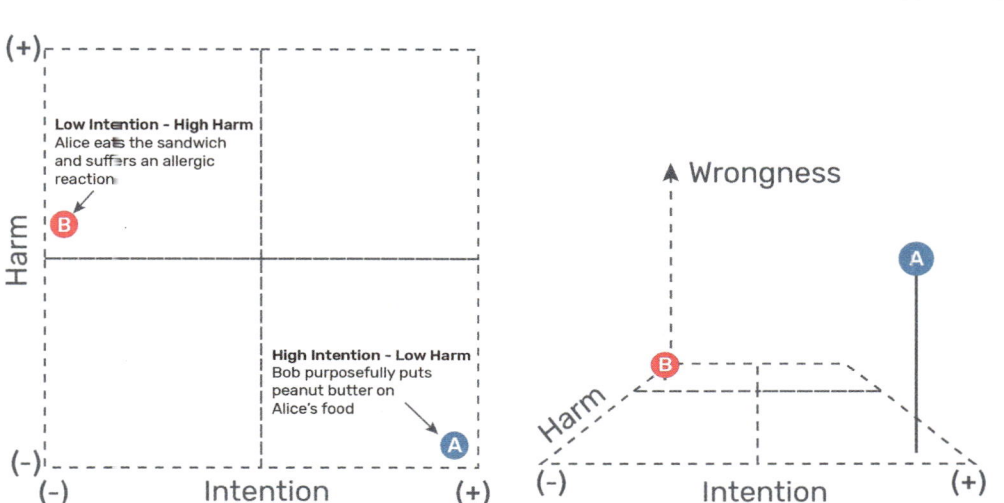

Figure 1.1

Moral space for the peanut butter scenarios.

the perceived level of intention I, the moral dimensions involved (H, F, L, A, and P), the characteristics c_i of the people—or machines—involved in the scenario, and the characteristics c_j of the person judging the scenario:

$$W = f(I,H,F,L,A,P,c_i,c_j).$$

We will explore this function empirically in chapter 6. One of the main questions posed there will be whether the function describing humans judging the actions of other humans (f_h) is different from the function describing humans judging the actions of machines (f_m). We will also discuss whether people with different demographic characteristics (c_j), such as gender, education, ethnicity, and so on, judge human and machine actions differently.

But should we expect any difference, or should we expect people's judgment of human actions to translate seamlessly to the actions of machines? For the time being, we should not get ahead of ourselves. In the next and final section of this chapter, we will describe the methodology that we use to collect our data. This will provide a guide to understand the figures and experiments presented in the following chapters.

How Humans Judge Machines

In this book, we explore dozens of scenarios comparing people's reactions to human and machine actions. *Scenarios* are short stories that describe an action that can have a positive or negative moral outcome. Each scenario was presented to different people as either the action of a human or a machine (AI). About 150 to 200 people evaluated each scenario in each condition (human or machine). We use the word *scenario* instead of *dilemma* because we are not asking subjects to tell us how *they* would behave, but rather to judge the behavior of the human or the machine. Also, some of these scenarios do not involve a dilemma per se; they may include accidents, transgressions, mistakes, or even situations in which a human or a machine corrects an unfair outcome.

To begin, consider the following scenario:

 A [driver/autonomous excavator] is digging up a site for a new building. Unbeknownst to the [driver/excavator], the site contains a grave. The [driver/ excavator] does not notice the grave and digs through it. Later, human remains are found.

In response to scenarios like this one, subjects were asked to answer a set of questions using a Likert-type scale. In this case, we used the following questions. Bold characters show the labels used to represent the answers to these questions in charts:

- Was the action **harmful**?
- Would you **hire** this driver for a similar position?
- Was the action **intentional**?
- Do you **like** the driver?
- How **morally** wrong or right was the driver's action?
- Do you agree that the driver should be **promoted** to a position with more responsibilities?
- Do you agree that the driver should be replaced with a robot or an algorithm? **[replace different]**
- Do you agree that the driver should be replaced by another person? **[replace same]**
- Do you think the driver is **responsible** for unearthing the grave?
- If you were in a **similar situation** as the driver, would you have done the same?

These questions were answered by subjects recruited online using Amazon Mechanical Turk (MTurk).‡ MTurk is an online crowdsourcing platform that has become a popular place to run social science experiments. While in principle, our results should be considered valid only for the specific people who participated in the MTurk exercise, in practice, various studies have shown that MTurk samples provide representations of the US population that are as valid as those obtained through commercial polling companies,[38] and are more representative than in-person convenience samples.[39] We leave the study of the same scenarios for non-US populations as a topic for future research.

To measure the moral dimensions associated with each scenario, we conducted a second data collection exercise in MTurk, where we asked people to associate words with each scenario. We provided people with four words per moral dimension (two positive and two negative), as shown in table 1.1, and asked them to pick the four words that best described each scenario—in order—from the list of twenty.

HARM	FAIRNESS	LOYALTY	AUTHORITY	PURITY
harmful (-)	unjust (-)	disloyal (-)	disobedient (-)	indecent (-)
violent (-)	discriminatory (-)	traitor (-)	defiant (-)	obscene (-)
caring (+)	fair (+)	devoted (+)	lawful (+)	decent (+)
protective (+)	impartial (+)	loyal (+)	respectful (+)	virtuous (+)

Table 1.1

Words used to associate scenarios to moral dimensions.

‡ The experimental procedure was approved by the IRB office at the Massachusetts Institute of Technology (MIT). COUHES Protocol # 1901642021.

For instance, if people associate a scenario with the words *discriminatory* or *unjust*, that tells us that this scenario involves the fairness dimension. If people associate a scenario with the words *indecent* and *obscene*, that tells us that this scenario touches on purity. The good thing about this technique is that it is nonbinary, meaning that we can use it to decompose a moral dilemma into multiple dimensions.

Figure 1.2 shows the moral dimensions associated with the excavator scenario presented earlier. Here, we show the fraction of times that people chose a word associated with each moral dimension. In this case, the scenario is associated strongly with purity (about 40 percent of word associations), and more mildly with harm and fairness (about 20 percent and 25 percent of word associations, respectively). This is reasonable because it describes the case of unearthing a dead body, considered a sacrilege by most cultures.

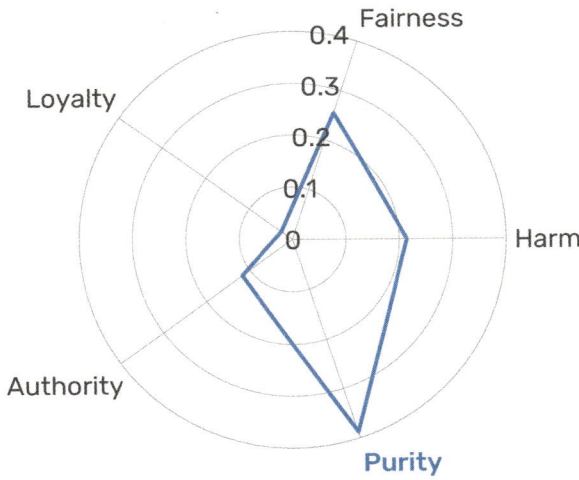

Figure 1.2

Moral dimensions associated with the excavator scenario.

In the next chapters, we will extend this exercise to multiple scenarios to create counterfactuals for the way in which humans judge machines. Figure 1.3 uses the excavator scenario to illustrate how we present our results. Here, the dots represent average values, and the error bars show 99 percent confidence intervals. Going forward, we use red to show data on humans judging machines, and blue to show data on humans judging humans. An easy way to remember this is to think: "Red is for robots."

Figure 1.3 shows that people rate the action of the autonomous excavator as more harmful and more morally wrong (lower values in the morality scale mean less moral). They also like the human more and are less inclined to want to promote machines. But how large are these differences? Are they just fluctuations, or are they meaningful? Here, we compare answers using both *p*-values and graphical statistical methods. *p*-values tell us the probability that the two answers are the same.

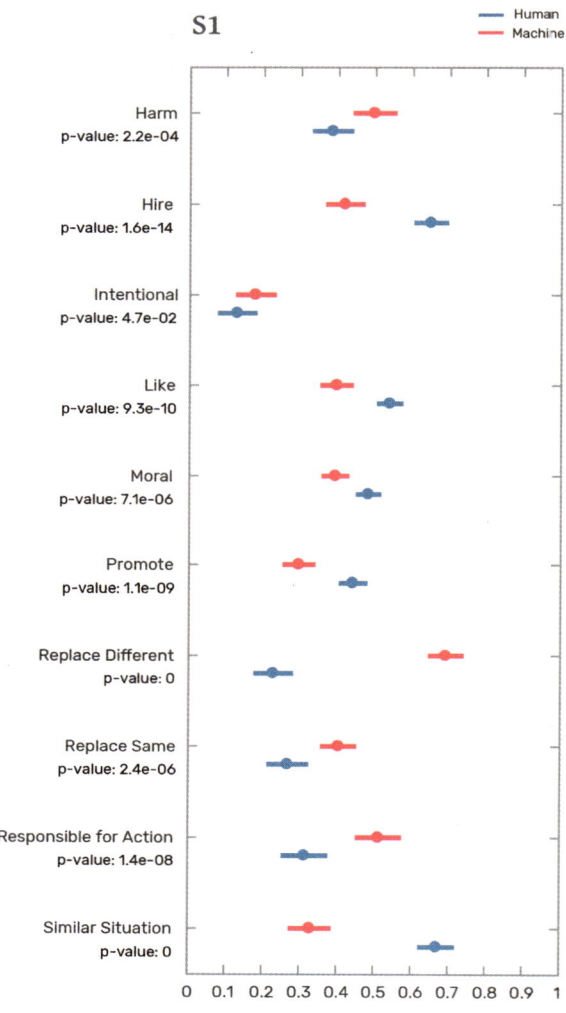

Figure 1.3

Judgments of excavator scenario.

When that probability is low (1 in 10,000 or 1 in a million), we can be quite certain that the two groups evaluated the scenarios differently. Yet, p-values do not tell the full story. While scholars have long used the concept of statistical significance and the idea of p-values to compare differences among groups, recently scientists[40] and statisticians[41] have stood against the practice of using p-values.

The critique is that using p-value thresholds (usually 1 in 50 and 1 in 100) as dichotomous measures of what is significant has created perverse incentives. Instead, these communities of scholars are advocating for the use of a more continuous approach to statistics. Here, we subscribe to this idea by including graphical methods to compare the data throughout the book. Graphical methods provide information that is hidden when using only p-values. For instance, in the excavator scenario, both "moral" and "replace same" have a similar p-value, but graphically behave differently (e.g., "moral" shows less difference and less variance).

In recent years, advances in machine learning have brought the idea of AI back into the limelight. Yet, we still have much to learn about how humans judge machines. In this chapter, we have introduced some basic AI concepts, such as the idea of strong and weak AI, as well as basic concepts from moral philosophy and moral psychology. In the next chapters, we will use these concepts to interpret experiments comparing people's reactions to scenarios involving humans and machines.

Unpacking the Ethics of AI

2

CHAPTER 2

In this chapter, we explore a number of experiments revealing people's attitudes toward artificial intelligence (AI). They compare people's reactions to humans and machines performing the same action. In each of these experiments, hundreds of subjects were randomly assigned to a treatment or control group. This means that the subjects who evaluated the AI actions did not see scenarios describing human actions, and vice versa. In the treatment condition, actions were performed by AI agents or robots, while in the control condition, the same actions were performed by a human. Otherwise, the scenarios were identical. By using a random assignment to either the treatment or the control group, we avoid any selection bias. For instance, if any of our subjects particularly liked or disliked technology, then they would have the same probability of being assigned to the treatment or control group.

In the next chapters, we use data from these experiments to compare people's attitudes toward AIs in a variety of scenarios. In this chapter, however, we will focus only on scenarios in four areas: involving risky, life-or-death decisions; lewd behavior; self-driving car accidents; and the desecration of national symbols. These four groups of scenarios will provide us with a quick overview of AI ethics and uncover an initial set of insights that we will continue to explore in the remainder of the book.

Risky Choices

Life is full of uncertainty. Yet we still need to make choices. In the future, AIs will also have to make choices in uncertain situations. But how will we judge them? Will we value risk-taking, or will we suppress the risk-taking qualities that we sometimes celebrate in humans?

Consider the following three versions of this moral dilemma:

A large tsunami is approaching a coastal town of 10,000 people, with potentially devastating consequences. The [politician/algorithm] responsible for the safety of the town can decide to evacuate everyone, with a 50 percent chance of success, or save 50 percent of the town, with 100 percent success.

S2 The [politician/algorithm] decides to save everyone, but the rescue effort fails. The town is devastated, and a large number of people die.

S3 The [politician/algorithm] decides to save everyone, and the rescue effort succeeds. Everyone is saved.

S4 The [politician/algorithm] decides to save 50 percent of the town.

All these scenarios are identical, in that they involve the same choice: a choice between a safe option that ensures 50 percent success and a risky option that has a 50 percent chance of success and a 50 percent chance of failure. While here we use a tsunami framing, we replicated this experiment with alternative framings (a forest fire and a hurricane, given in scenarios A1–A6 in the appendix) and obtained similar results.

In all three scenarios, 50 percent of people survive (on average). But while the three scenarios have the same expected outcome, they differ in what actually occurs. In the first scenario, the risky choice results in failure, and many people die. In the second scenario, the risky choice results in success, and everyone lives. In the third scenario, the compromise is chosen, and half of the people are saved.

About 150 to 200 subjects, who saw only one of the six conditions (risky success, risky failure, or compromise, as either the action of a human or a machine), evaluated each scenario. Having separate groups of subjects judge each condition reduces the risk of contaminating the results from exposure to similar cases.

But how did people judge the actions of AIs and humans?

Figure 2.1 shows average answers with their corresponding 99 percent confidence intervals. We can quickly see large differences in the risky scenarios (S2 and S3). In the case in which the action involves taking a risk and failing, people evaluate the risk-taking politician much more positively than the risk-taking algorithm. They report that they like the politician more, and they consider the politician's decision as more morally correct. They also consider the action of the algorithm as more harmful. In addition, people identify more with the decision-making of the politician because they are more likely to report that they would have done the same when the risky choice is presented as a human action. Surprisingly, people see both the action of the algorithm and that of the politician as equally intentional.

On the contrary, in the scenario where the risk resulted in success (S3), people see the politician's action as more intentional. In this situation, they evaluate the politician much more positively than the algorithm. They like the politician more, consider their action as more morally correct, and are more likely to want to hire or promote them.

In the compromise scenario, however, we see almost no difference. People see the action of the politician as more intentional, but they rate the politician and the algorithm equally in terms of harm and moral judgment. We also do not observe significant differences in people's willingness to hire or promote the politician or the algorithm, and they report liking both the same.

But why do we observe such marked differences?

On the one hand, these results agree with previous research showing that people quickly lose confidence in algorithms after seeing them err, a phenomenon known as *algorithm aversion*.[1] On the other hand, people may be using different mental models to judge the actions of the politician and the algorithm. Consider the concepts of moral agency and moral status introduced in chapter 1. In the tsunami scenario, a human decision-maker (the politician) is a moral agent who is expected to acknowledge the moral status of everyone. Hence, they are expected to try to save all citizens, even if this is risky. Thus, when the agent fails, they are still evaluated positively because they tried to do the "right" thing. Moral agents have a metaphorical heart, and they are evaluated based on their ability to act accordingly. A machine in the same situation, however, does not enjoy the same benefit of the doubt. A machine that tries to save everyone, and fails, may not be seen as a moral agent trying to do the right thing, but rather as a defective system that erred because of its limited capacities. In simple words, in the context of a moral dilemma, people may expect machines to be rational and people to be human.

But are these results generalizable? Are we less forgiving of AIs
when they make the same mistakes
as humans, or is this true only for
some types of mistakes? To explore these questions, let's move on to
the next group of scenarios.

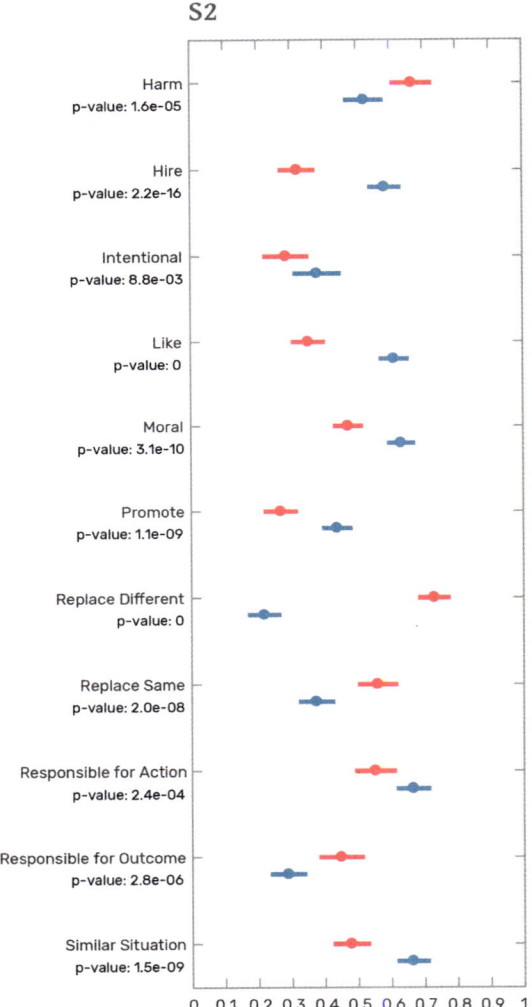

Figure 2.1

Participant reactions to three tsunami
scenarios (S2,S3,S4).

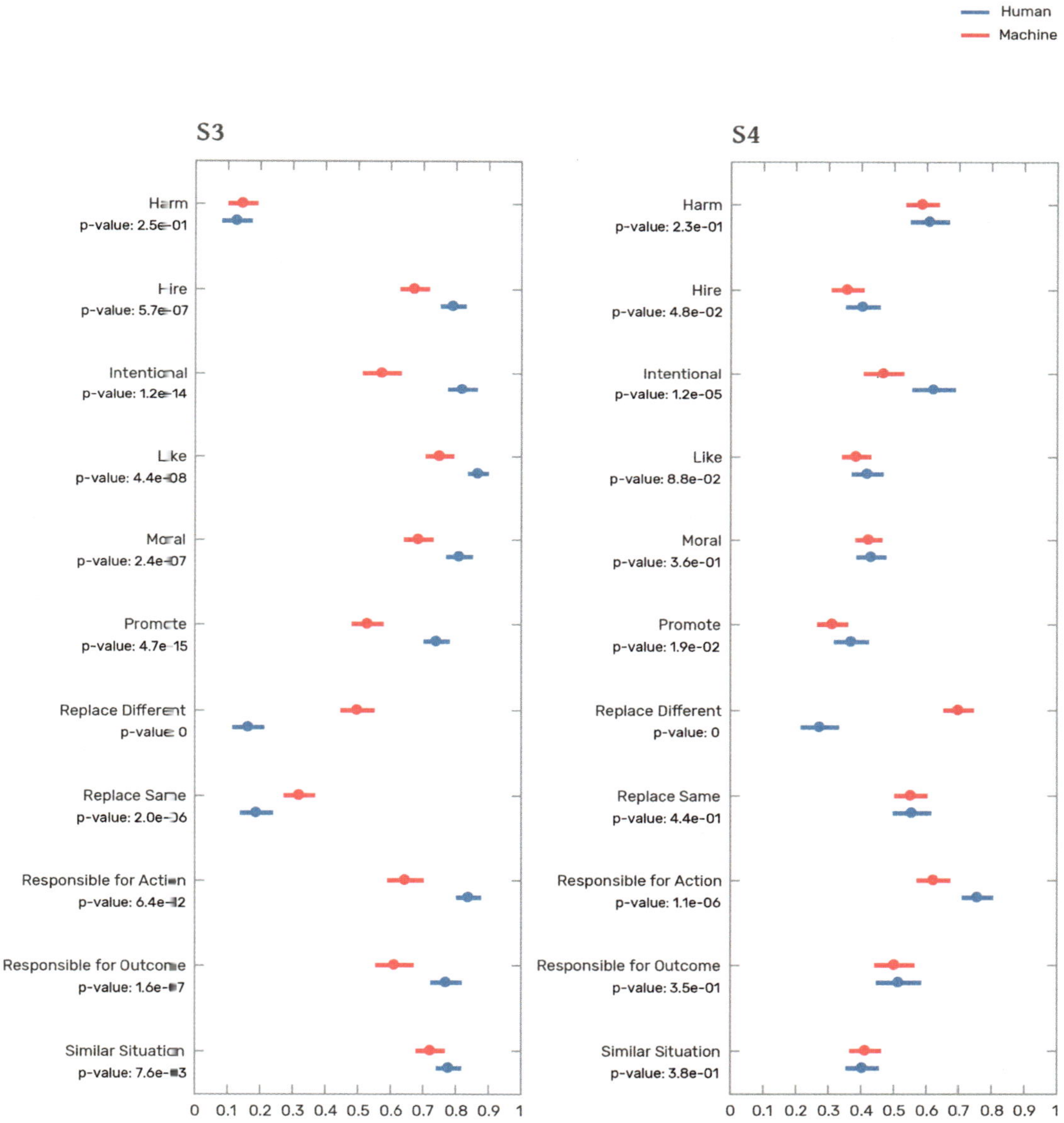

UNPACKING THE ETHICS OF AI

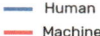
Human
Machine

S3

	Human	Machine
Harm p-value: 2.5e-01		
Hire p-value: 5.7e-07		
Intentional p-value: 1.2e-14		
Like p-value: 4.4e-08		
Moral p-value: 2.4e-07		
Promote p-value: 4.7e-15		
Replace Different p-value: 0		
Replace Same p-value: 2.0e-06		
Responsible for Action p-value: 6.4e-12		
Responsible for Outcome p-value: 1.6e-07		
Similar Situation p-value: 7.6e-03		

0 0.1 0.2 0.3 0.4 0.5 0.6 0.7 0.8 0.9 1

S4

	Human	Machine
Harm p-value: 2.3e-01		
Hire p-value: 4.8e-02		
Intentional p-value: 1.2e-05		
Like p-value: 8.8e-02		
Moral p-value: 3.6e-01		
Promote p-value: 1.9e-02		
Replace Different p-value: 0		
Replace Same p-value: 4.4e-01		
Responsible for Action p-value: 1.1e-06		
Responsible for Outcome p-value: 3.5e-01		
Similar Situation p-value: 3.8e-01		

0 0.1 0.2 0.3 0.4 0.5 0.6 0.7 0.8 0.9 1

Trouble at the Theater

In principle, creative tasks seem to be uniquely human. In practice, however, weak forms of AI are becoming important sources of creativity.[2] AIs now can generate synthetic photographs, text, and videos using techniques such as Generative Adversarial Networks (GANs).[3]

The rise of artificial creativity is motivating various debates. On the one hand, the ability of AIs to create content has fueled an active debate about copyright, with arguments in favor and against the idea of assigning copyrights to algorithms or their creators.[4] On the other hand, the use of *deep fake* videos,[5] which can be used to put words in someone else's mouth, is raising concerns about the veracity of online content and the potential manipulation of political campaigns. Deep fakes can be used to create content resembling the appearance and voice of famous politicians, as well as blending someone's face onto pornographic material. As a result, the creative and media industries are now in a digital arms race between the tools that make synthetic content and those designed to detect it.[6]

But the creativity of AI systems is not only limited to imagery. The people working on creative AI are also exploring the creation of text. From tweeting bots to fake news articles, AIs are increasingly becoming a central part of our creative world. Platforms such as Literai, Botnik, or Shelley AI,* gather communities of people who use AI to create literary content.

Generative AIs have already become commonplace in the production of simple, data-driven news stories, like those related to weather or stock market news.[7] More recently, however, these efforts have moved to more complex literary creations.

---

* See https://www.literai.com/, http://botnik.org, http://shelley.ai.

In their literary incarnations, many of these efforts can capture the voice, tone, and rhythm of famous authors. But at the same time, these tools can fail to produce the narrative coherence expected from a literary work.[†] For example, here are two passages from a *Harry Potter* chapter created by Botnik:

"The castle grounds snarled with a wave of magically magnified wind. The sky outside was a great black ceiling, which was full of blood. The only sounds drifting from Hagrid's hut were the disdainful shrieks of his own furniture."

This passage is quite good, but this is not true of all passages:

"'Voldemort, you're a very bad and mean wizard' Harry savagely said. Hermione nodded encouragingly. The tall Death Eater was wearing a shirt that said 'Hermione Has Forgotten How to Dance,' so Hermione dipped his face in mud."

As the capacity of these technologies continues to improve,[8] we will encounter a world where AIs probably will not be involved in creative decisions, but they nevertheless will become part of the creative teams providing the options that artists and creative directors use as input. Like spoiled teenagers, our creative future may involve choosing among countless options generated by algorithms that are programmed to seek our approval. This revolution not only will affect visual arts and literature, but also will reach other domains, like the use of creative AIs to create new recipes,[9] generate data visualizations,[10] and compose music.[11]

But how will we judge our new creative companions? Will we give them a seat at the writer's table? Will we allow them to be as expressive as they can be? Or will we censor them relentlessly?

[†] The algorithms are unable to communicate a larger idea or make a point with their stories, as a human would do. They are stuck in short-term correlations of words instead of generating long-term correlations of concepts.

In this section, we explore some of the ethical questions involving creative uses of AI. How do people judge AIs that are lewd, disrespectful, or blasphemous? How tolerant are we toward creative AIs? Do we punish them more severely than humans who have committed the same transgressions?

To begin, consider the following three marketing scenarios:

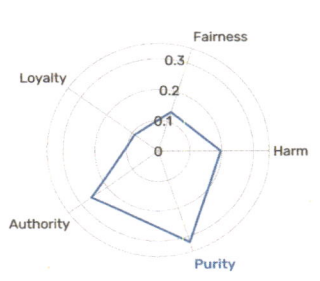 

S5   A well-known clothing company wants to create a high-impact commercial. It decides to hire a new [marketeer/AI marketing system] to design an image that combines rivalry and love. The results are the images above, which cause shock and outrage among some members of the public.

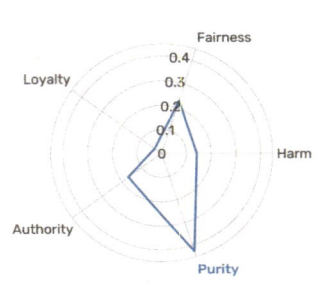 

S6   A public transportation company wants to create a funny commercial. It decides to commission an advertisement from a(n) [marketeer/AI marketing system] that uses a play on the word *riding*. The resulting ad, pictured above, causes shock and outrage among members of the public.

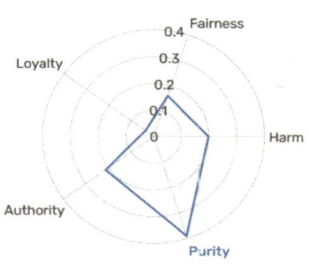

S7 A fashion company wants a new advertisement that illustrates addiction to clothes and fashion. The company employs a(n) [marketeer/AI marketing system] to design an ad that uses the concept of addiction as its main message. The resulting advertisement, pictured above, causes shock and outrage among members of the public.

Our findings for these three scenarios are presented in figure 2.2. These scenarios show a very similar pattern of results. People dislike the human and the algorithm similarly. They also don't see either as more morally right. Not surprisingly, they assign more intention to the human than the AI.

What is interesting about these scenarios is that they include explicit questions about the assignment of responsibility up the hierarchy. We asked subjects: Who is more responsible for the images (the marketeer or the company)? And who should respond to the public (the marketeer or the company)? Here, we find important differences. In both cases, we see responsibility move up the hierarchy when the algorithm is involved in the creative process. This suggests that the introduction of AI may end up centralizing responsibilities up the chain of command.

While simple, the observation that responsibility moves up the hierarchy when using AI is important because one of the reasons why people delegate work in an organization is to pass responsibility to others. In case of failure, delegation provides a "firewall" of sorts because blame can be passed from the management team to those involved in the execution of a task. In cases of success, those in charge can still take credit for the work of those whom they manage. Using AI eliminates the firewall, and hence can create a disincentive for the adoption of AI among risk-averse management teams.

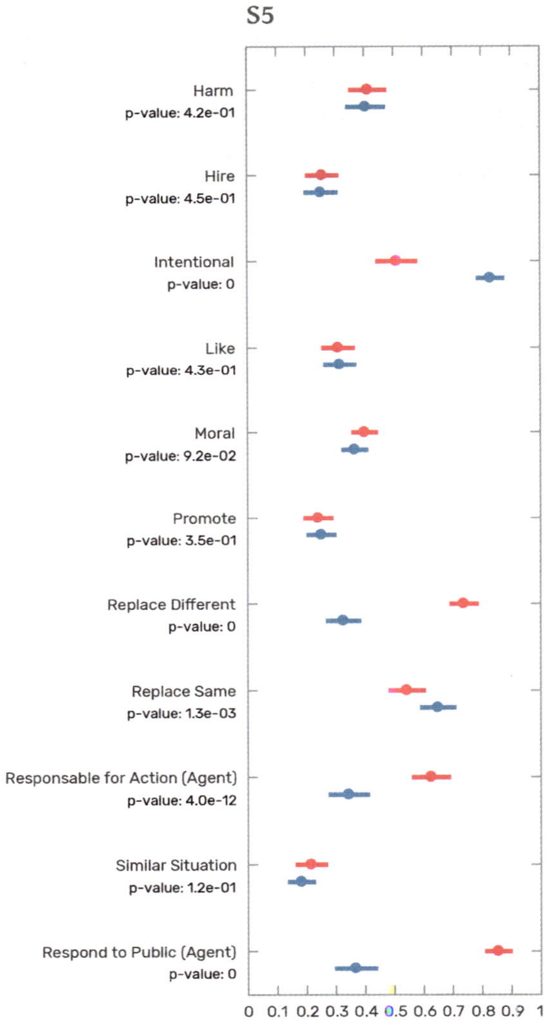

Figure 2.2

Participant reactions to three marketing scenarios (S5,S6,S7).

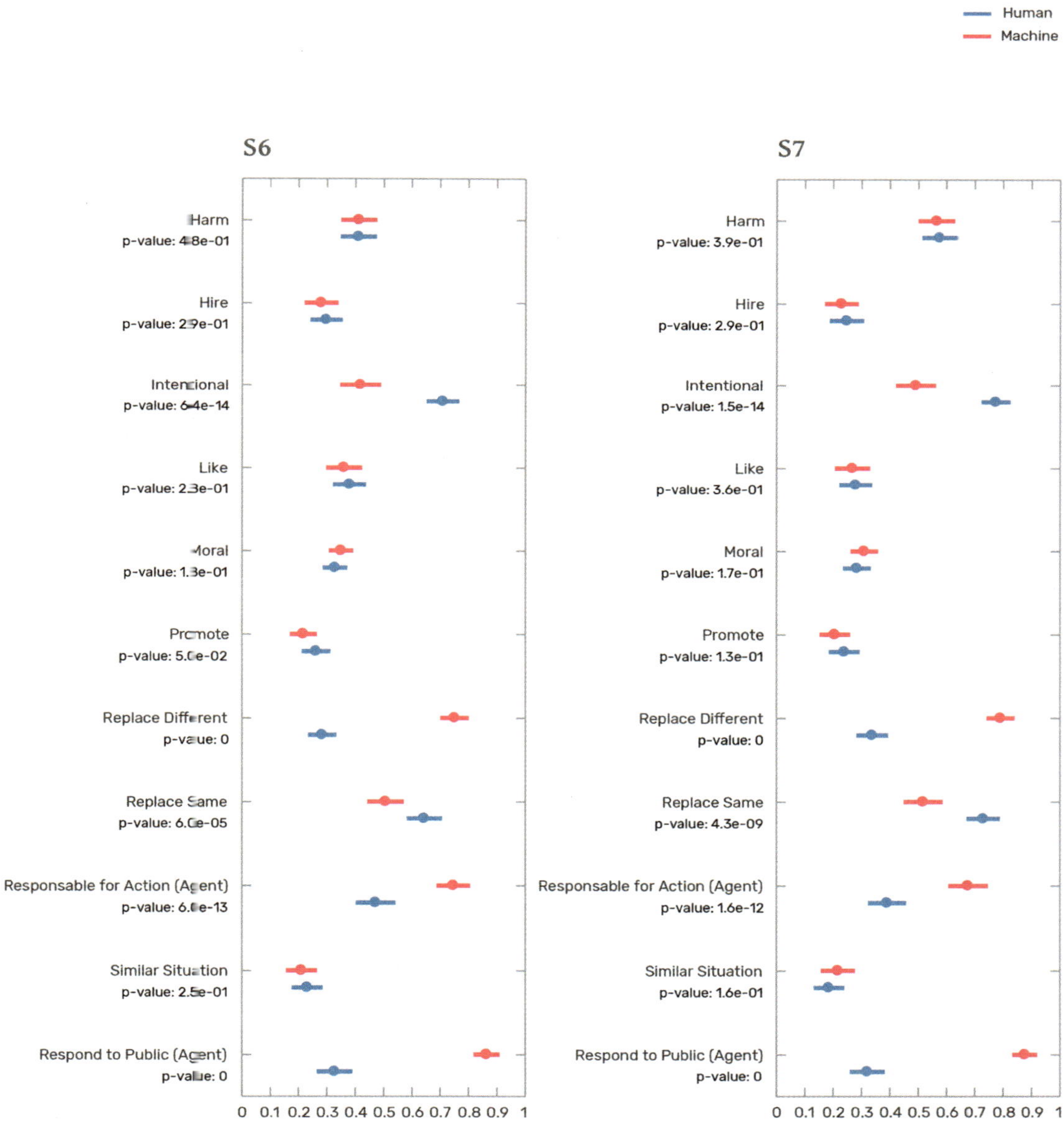

UNPACKING THE ETHICS OF AI

Human
Machine

S6

Harm	
p-value: 4.8e-01	
Hire	
p-value: 2.9e-01	
Intentional	
p-value: 6.4e-14	
Like	
p-value: 2.3e-01	
Moral	
p-value: 1.3e-01	
Promote	
p-value: 5.0e-02	
Replace Different	
p-value: 0	
Replace Same	
p-value: 6.0e-05	
Responsable for Action (Agent)	
p-value: 6.0e-13	
Similar Situation	
p-value: 2.5e-01	
Respond to Public (Agent)	
p-value: 0	

0 0.1 0.2 0.3 0.4 0.5 0.6 0.7 0.8 0.9 1

S7

Harm	
p-value: 3.9e-01	
Hire	
p-value: 2.9e-01	
Intentional	
p-value: 1.5e-14	
Like	
p-value: 3.6e-01	
Moral	
p-value: 1.7e-01	
Promote	
p-value: 1.3e-01	
Replace Different	
p-value: 0	
Replace Same	
p-value: 4.3e-09	
Responsable for Action (Agent)	
p-value: 1.6e-12	
Similar Situation	
p-value: 1.6e-01	
Respond to Public (Agent)	
p-value: 0	

0 0.1 0.2 0.3 0.4 0.5 0.6 0.7 0.8 0.9 1

Next, we look at three additional examples in the creative industries: one involving a plagiarizing songwriter, one involving a blasphemous comedian, and another describing a lewd playwright:

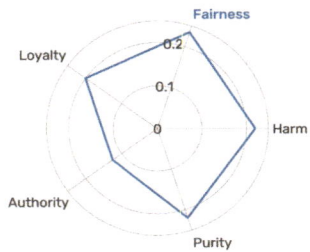

S8 A record label hires a(n) [songwriter/AI songwriter] to write lyrics for famous musicians. The [songwriter/AI songwriter] has written lyrics for dozens of songs in the past year. However, a journalist later discovers that the [songwriter/AI songwriter] has been plagiarizing lyrics from lesser-known artists. Many artists are outraged when they learn about the news.

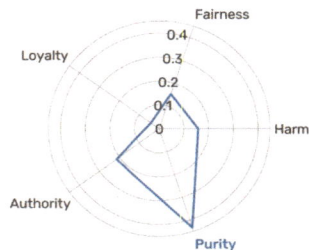

S9 A TV studio decides to employ a(n) [comedian/AI comedy software] to write sketches for a new show. The [comedian/AI] writes a sketch in which God is sucking the penis of the devil. The piece is controversial, and many people are deeply offended.

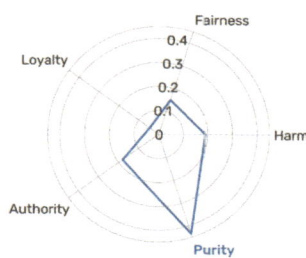

S10 A theater decides to hire a new [artist/AI algorithm] to prepare a performance art piece. In the piece, actors have to act like animals for 30 minutes, including crawling around naked and urinating onstage. Some members of the audience are disgusted and offended.

The case of the songwriter is interesting because AIs rely on massive training data sets, which can give AIs a herdlike property. Because AIs learn from examples, creative outcomes that reuse parts of those examples could result in plagiarism.[12] The cases of the comedy sketch and of the performance art piece, on the other hand, are examples of creative outcomes that break social norms associated with the moral dimension of purity. The comedy sketch can be perceived as both lewd and blasphemous, whereas the performance art piece could be considered by some as grotesque or lewd, but not blasphemous.

The results for these three cases are presented in figure 2.3. In these cases, we find that the action of the human is seen as more intentional than that of the AI. The responsibility also moves up the command chain, confirming what we found in the advertisement examples. Also, as in all the previous cases, people are eager to replace AIs with humans.

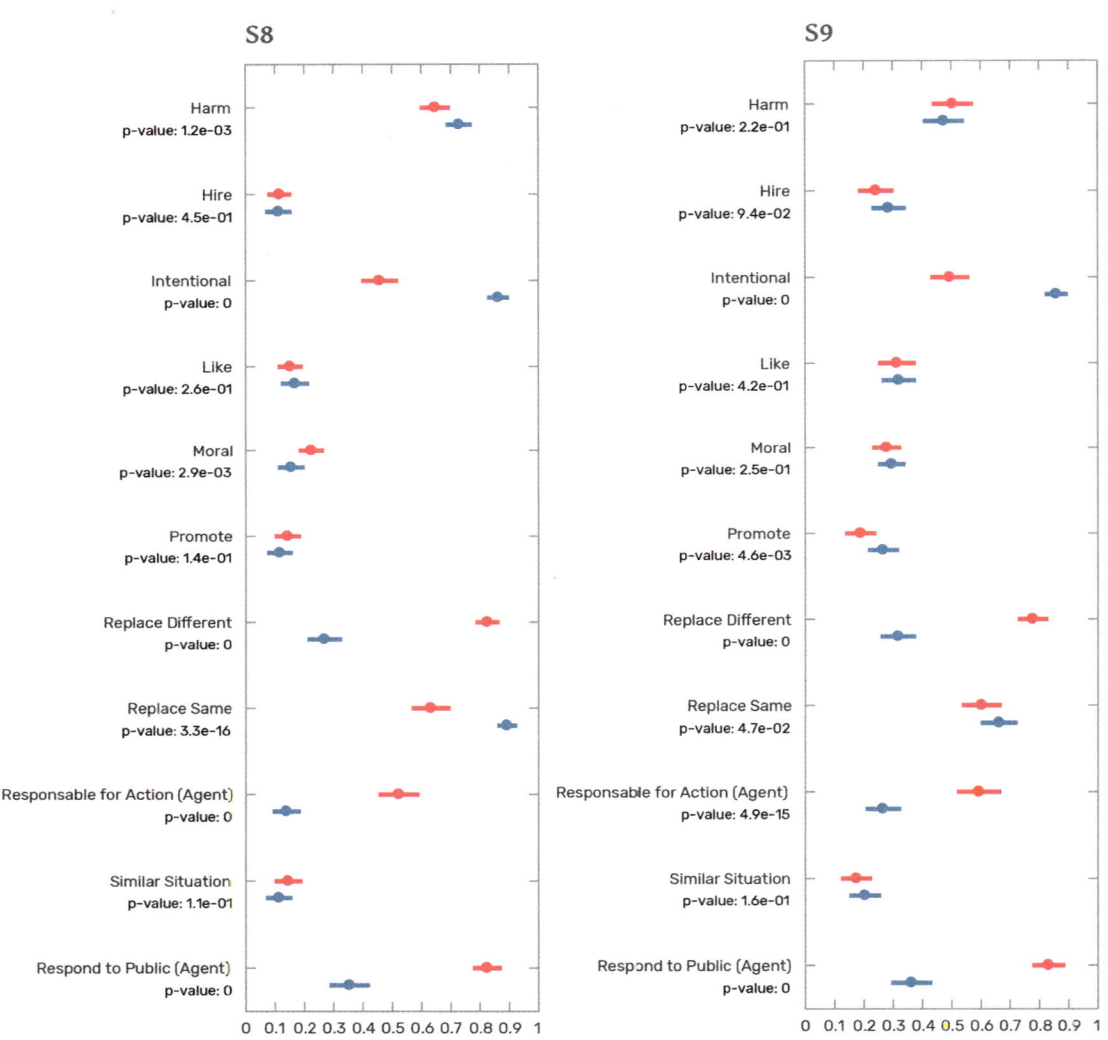

Figure 2.3

Participant reactions to three creative industry scenarios
(S8,S9,S10).

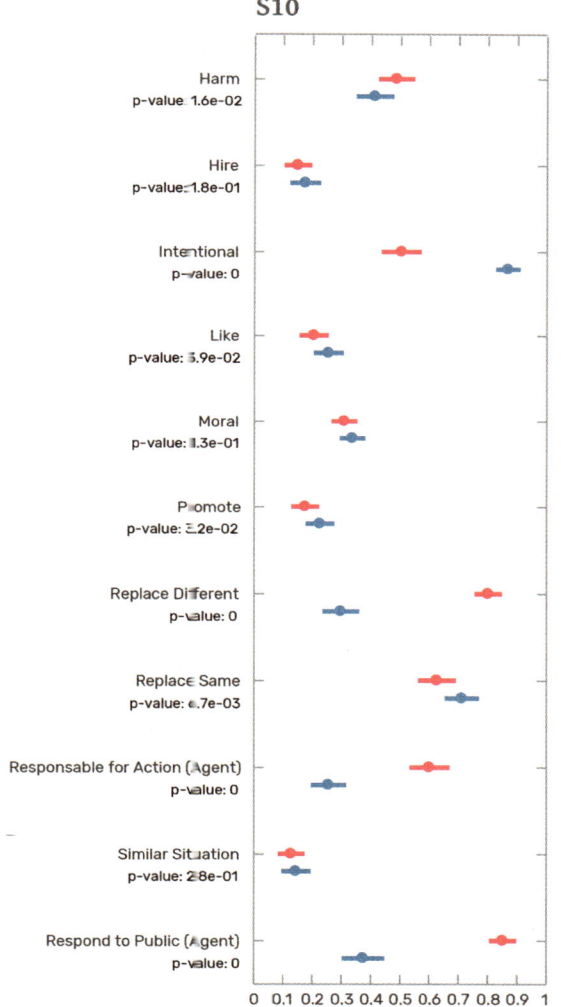

Other than that, we don't observe big differences in people's judgments of AI or humans except in the plagiarism scenario, where people judge the action of the human as slightly less moral. This is interesting because unlike the TV studio and the theater scenarios, which involve the moral dimensions of purity, the plagiarism scenario is heavier in the moral dimension of fairness. This suggests that people may be less forgiving of other humans in scenarios that involve unfair behavior, suggesting that the moral dimension modulates whether the human or the machine is judged more harshly.

But how are humans and machines judged in scenarios involving accidents? In the next section, we explore questions involving traffic accidents that will help us revise our intuition about the relationship between AI, humans, and intentionality.

Watch Out!

Self-driving cars, or autonomous vehicles, are one of the examples of automation that is on everyone's mind.[13] Yet self-driving technologies are not only disrupting the passenger vehicle sector. In the last decade, these technologies have been deployed or tested in a variety of industries, from freight transportation to mining.

In 2005, for instance, Komatsu, a Japanese heavy machinery company; and Codelco, Chile's state-owned mining company, began piloting autonomous trucks in an active mine.[14] These trucks were deployed in 2008, in Codelco's Gaby mine and in an Australian mine operated by Rio Tinto. Nowadays, self-driving trucks, or *autonomous hauling systems* (*AHSs*), as they are called in the mining industry, are an increasingly common sight in mines across the world.

During recent years, the rise of autonomous vehicles has escaped the controlled environments of mining operations. Self-driving freight convoys have completed thousands of kilometers[15] in Europe, and self-driving cars have completed millions of miles in the US.[16]

In recent years, a fertile stream of literature in AI ethics has focused on self-driving vehicles.[17] Scholars have studied the moral preferences that people would like to endow autonomous cars with[18] and how these preferences vary across the globe.[19] This research shows that people would refrain from buying self-sacrificing cars, although they would like other people to do so.[20] Further, this research has argued that some of the main roadblocks limiting the adoption of self-driving cars are psychological[21] rather than computational, and they include overreactions to autonomous vehicle accidents and the opacity of the autonomous decision-making process. In fact, despite much enthusiasm for the technology, people seem to be cautious about autonomous vehicles. A recent survey in the US found that three-quarters of Americans are afraid of riding in a self-driving car.[22]

But the issue with autonomous vehicles is that they don't fully eliminate accidents. So a question that remains is: How do we judge self-driving cars when they are involved in the same accidents as humans?

Here, we explore four scenarios to contribute to this growing literature:

   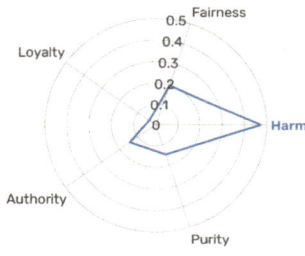

S11 On a sunny spring day, a [driver/driverless car] working for a supermarket chain accidentally runs over a pedestrian who runs in front of the vehicle. The pedestrian is hurt and is taken to the hospital.

   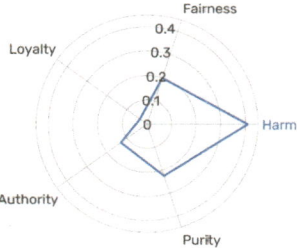

S12 On a sunny spring day, a [driver/driverless car] working for a supermarket chain accidentally runs over a dog that jumps in front of the vehicle. The dog is hurt and is taken to the veterinarian.

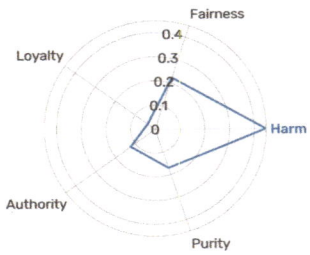

S13 On a cold and windy day, a [driver/driverless car] working for a supermarket chain swerves to avoid a falling tree. By swerving, the [driver/driverless car] loses control of the vehicle, leading to an accident that seriously injures a pedestrian on the sidewalk.

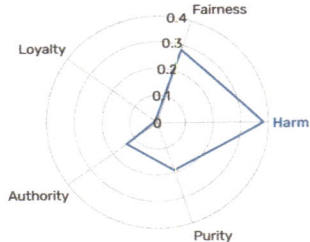

S14 On a cold and windy day, a [driver/driverless car] working for a supermarket chain swerves to avoid a falling tree. By swerving, the [driver/driverless car] loses control of the vehicle, leading to an accident that seriously injures a dog on the sidewalk.

These four scenarios can be grouped in two ways. First, when it comes to the victim, two scenarios involve a pedestrian and two a dog. This helps us vary the level of severity of the accident (as dogs have a lower moral status than humans). Also, the first two scenarios involve an accident in which a pedestrian or a dog jumps in front of the car. The second two scenarios involve a case in which the accident is triggered by an exogenous event (a falling tree), which causes the human or autonomous driver to lose control of the vehicle.

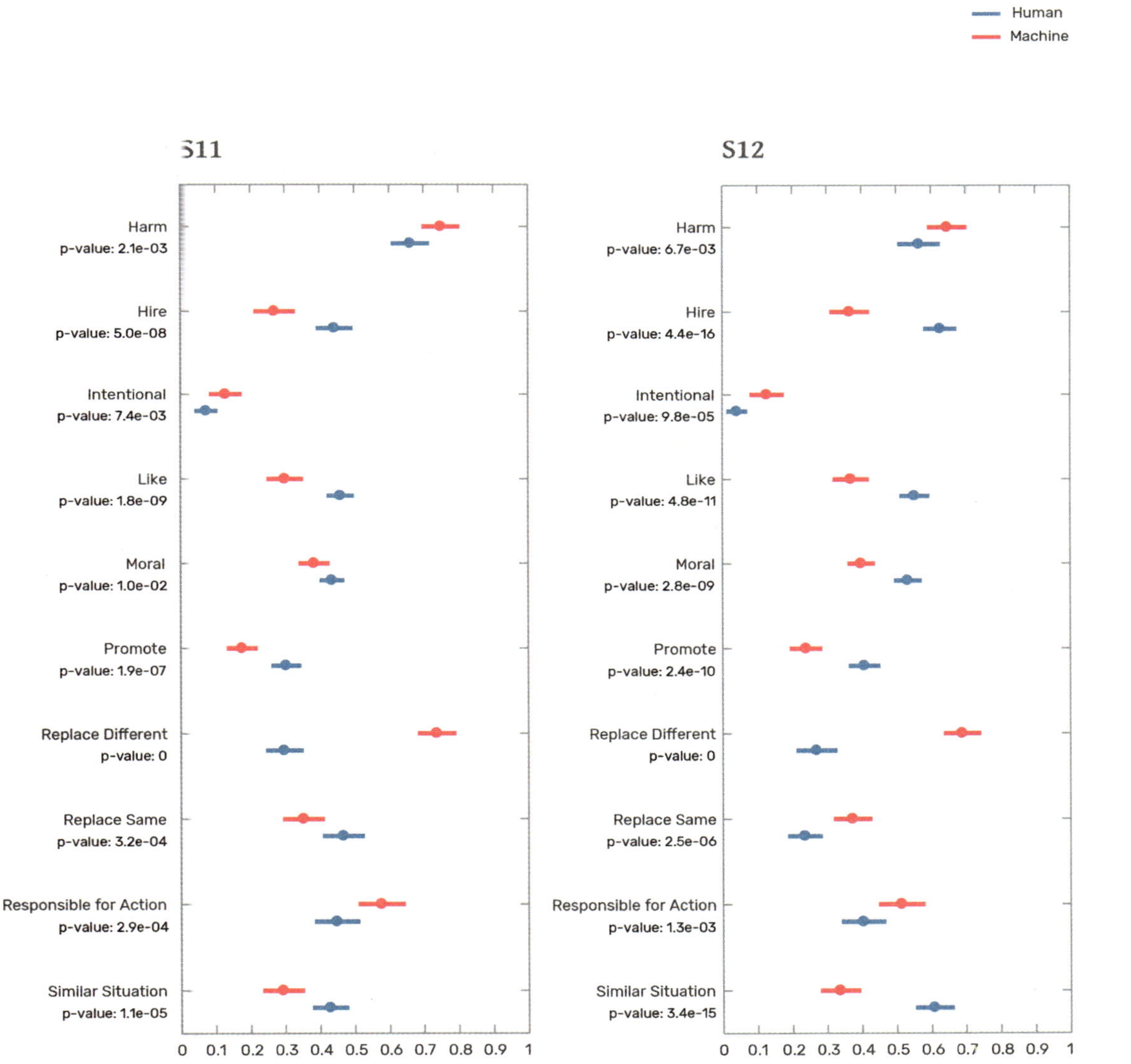

Figure 2.4

Participant reactions to four accident scenarios
(S11,S12,S13,S14)

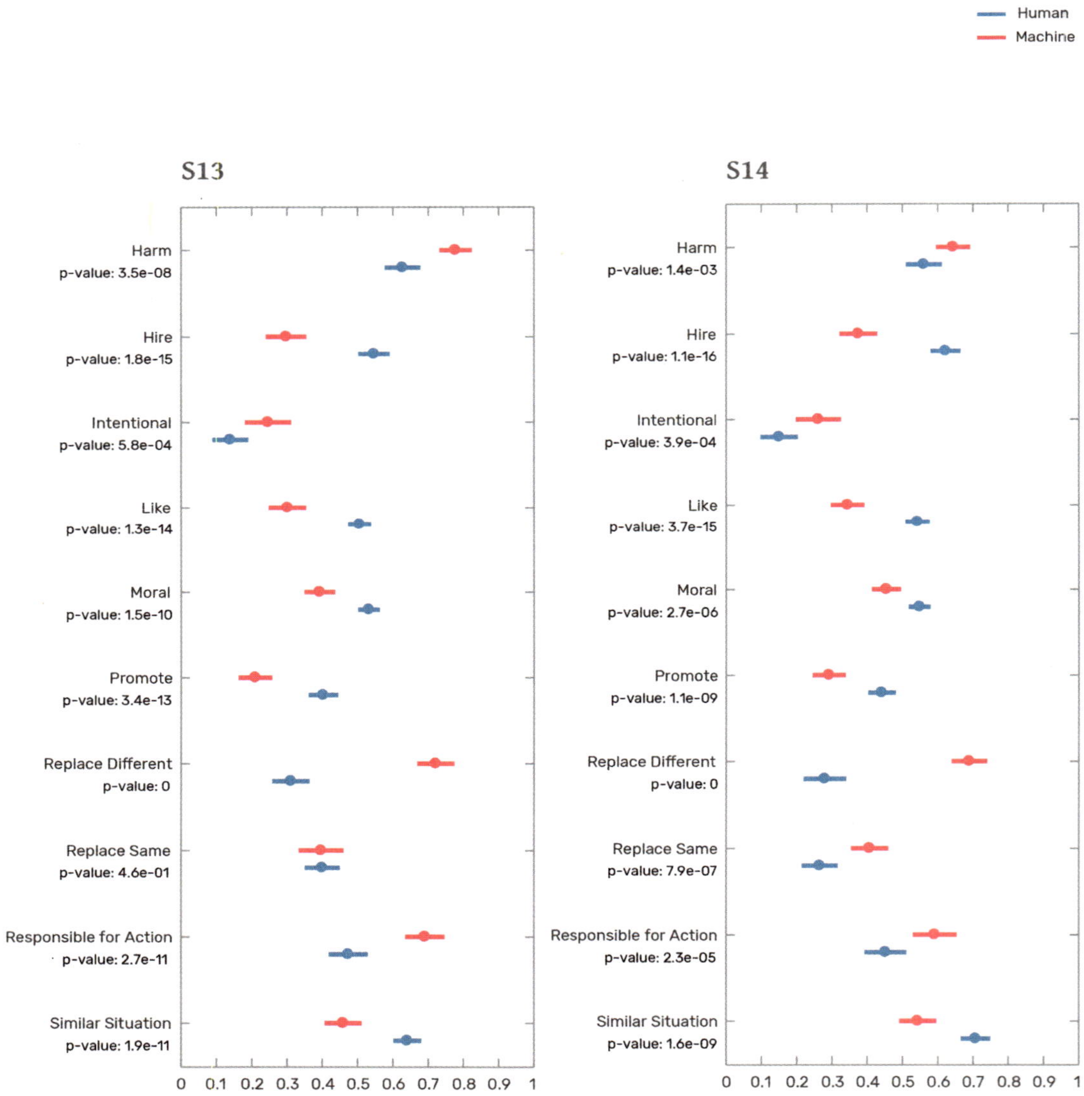

S13

Harm p-value: 3.5e-08	
Hire p-value: 1.8e-15	
Intentional p-value: 5.8e-04	
Like p-value: 1.3e-14	
Moral p-value: 1.5e-10	
Promote p-value: 3.4e-13	
Replace Different p-value: 0	
Replace Same p-value: 4.6e-01	
Responsible for Action p-value: 2.7e-11	
Similar Situation p-value: 1.9e-11	

0 0.1 0.2 0.3 0.4 0.5 0.6 0.7 0.8 0.9 1

S14

Harm p-value: 1.4e-03	
Hire p-value: 1.1e-16	
Intentional p-value: 3.9e-04	
Like p-value: 3.7e-15	
Moral p-value: 2.7e-06	
Promote p-value: 1.1e-09	
Replace Different p-value: 0	
Replace Same p-value: 7.9e-07	
Responsible for Action p-value: 2.3e-05	
Similar Situation p-value: 1.6e-09	

0 0.1 0.2 0.3 0.4 0.5 0.6 0.7 0.8 0.9 1

Legend: Human, Machine

Together, the four cases reveal some interesting patterns (figure 2.4). First, we observe that the accidents are seen as slightly more harmful when they involve an autonomous vehicle. This difference is mild in most cases, but it is particularly strong in the windy scenario involving a human victim (S13). We also observe that people are more likely to report that they would have done the same when the accident involves a human driver, meaning that they can more easily put themselves in the shoes of the human. This is true in all four cases here. People also evaluate the human driver more positively, reporting to like the driver more and seeing their action as more morally correct. What is surprising in these scenarios is that we observe a slight tendency for people to judge the action of the autonomous car as *more* intentional than that of the human. This tendency is not very strong, but it is interesting because it suggests that humans may be willing to forgive another human involved in an accident more than they would be willing to forgive a robot. These results appear to run counter to recent work showing that drivers are blamed more than autonomous vehicles in traffic accidents,[23] but this is not necessarily the case because in our experiments, accidents are not attributed to mistakes,[24] but to exogenous reasons.

So far, we have looked at cases in which humans and machines are judged similarly and where humans are judged more positively than machines. We have encountered only one case in which humans were judged more harshly (plagiarism). But are there more cases in which people are less forgiving to humans? In the next and final section of this chapter, we explore a different type of moral dilemma: those that do not involve harm, plagiarism, or lewd behavior, but rather offenses to national symbols. Will machines finally get a break in such cases?

Red Flags

In 2006, the US Senate voted on what could have become the Twenty-Eighth Amendment to the Constitution. The "flag-burning" amendment, as it was popularly known, was designed to prohibit the desecration of the US flag, especially by burning. The amendment was controversial, among other reasons, because the Supreme Court had already ruled on that issue in 1989. In *Texas v. Johnson*, the Supreme Court voted 5–4 that it was legal to burn a US flag because doing so was an act of communication protected by the First Amendment (free speech). Nevertheless, the amendment was approved by the House of Representatives and lost in the Senate by only one vote.[25] This all goes to show that when it comes to national symbols, people make strong moral judgments about the way in which others treat them. But what about flag-burning robots?

In this section, we explore four moral dilemmas involving humans and machines desecrating national symbols (i.e., flags and anthems). Consider these four scenarios:

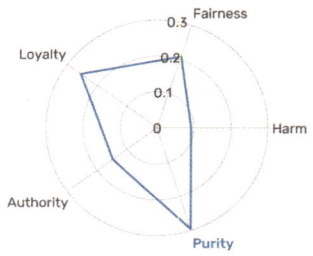   

S15 A family has a [cleaner/robot] in charge of cleaning their house. One day, the family finds that the [cleaner/robot] used an old national flag to clean the bathroom floor and then threw it away.

   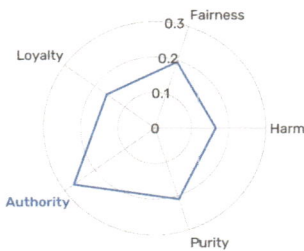

S16 During a major sporting event, the [operator/algorithm] running the public announcement system interrupts the national anthem to notify the crowd about a car that is poorly parked and is about to be towed.

   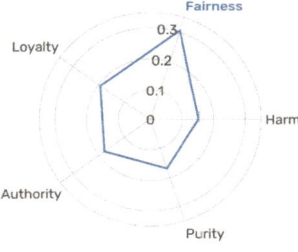

S17 In an international sporting event, the [operator/algorithm] running the public announcement system plays the wrong national anthem for one of the two teams. The fans in the station are baffled and annoyed.

   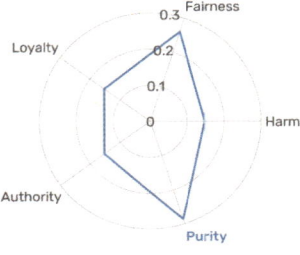

S18 A demolition crew, composed of [construction workers and heavy machinery/ autonomous heavy machinery],is tasked with tearing down an old public school that is scheduled for reconstruction. During the demolition process, the crew fails to notice that the American flag is still waving on the flagpole. The flag is shredded by the heavy machinery and is buried in the rubble.

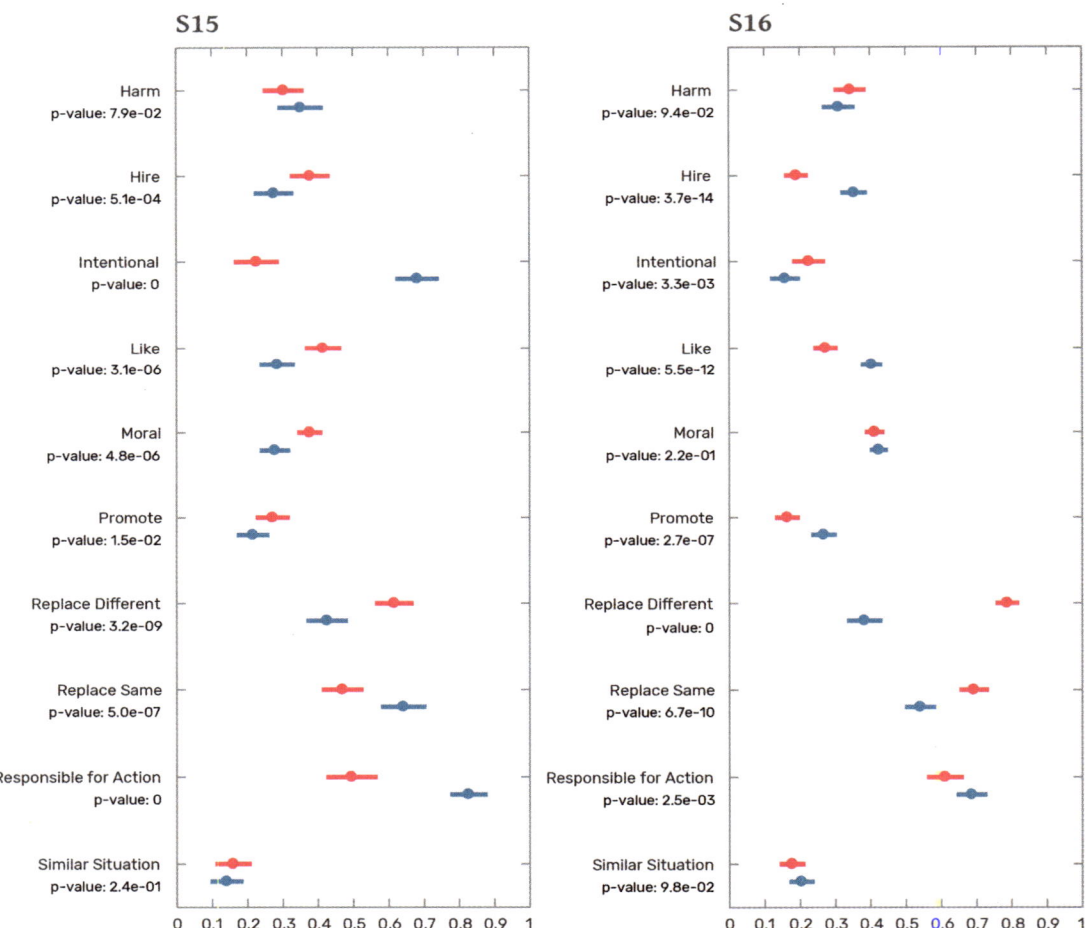

Figure 2.5

Participant reactions to four national symbol scenarios
(S15,S16,S17,S18).

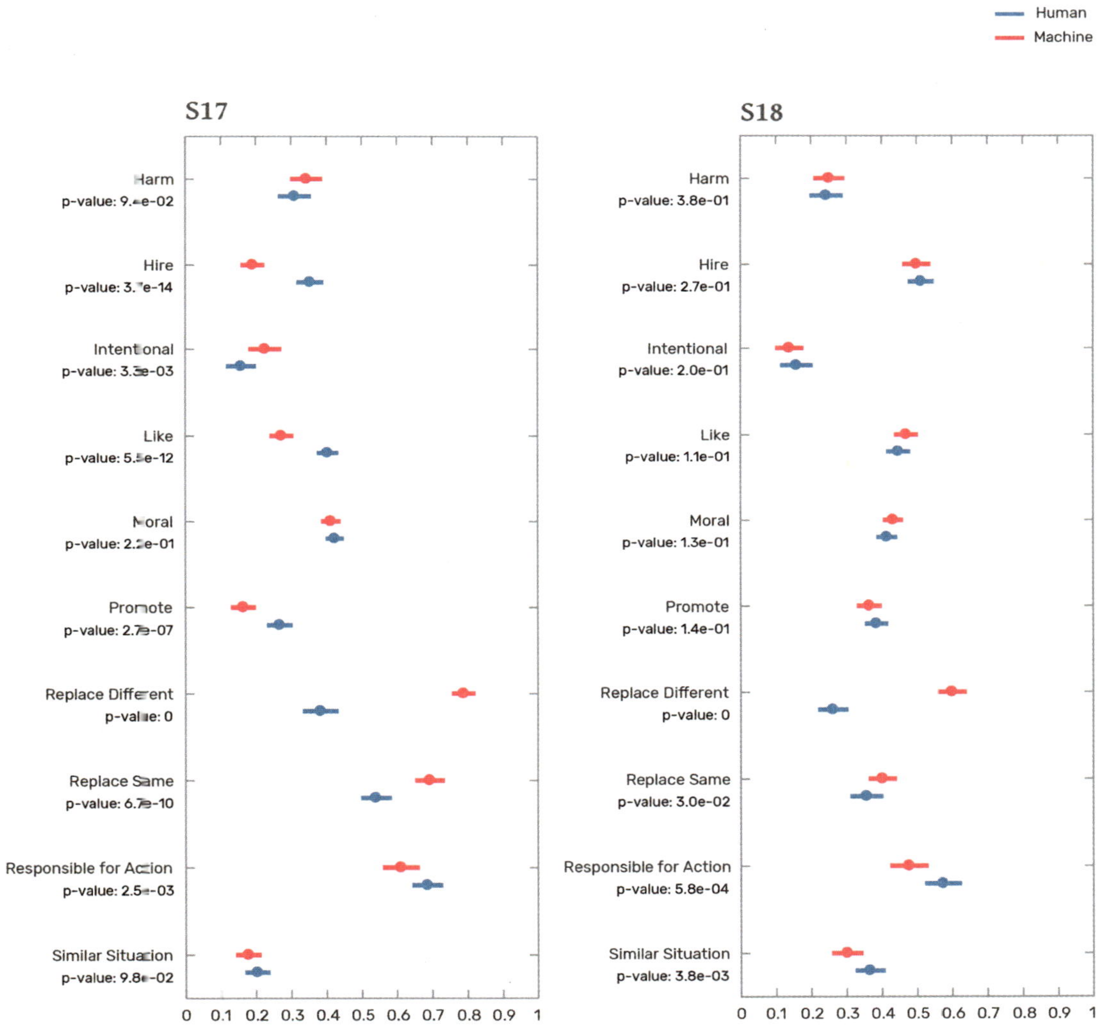
Human
Machine
S17
Harm
Hire
Intentional
Like
Moral
Promote
Replace Different
p-value: 0
Replace Same
Responsible for Action
Similar Situation
0 0.1 0.2 0.3 0.4 0.5 0.6 0.7 0.8 0.9 1
S18
Harm
p-value: 3.8e-01
Hire
p-value: 2.7e-01
Intentional
p-value: 2.0e-01
Like
p-value: 1.1e-01
Moral
p-value: 1.3e-01
Promote
p-value: 1.4e-01
Replace Different
p-value: 0
Replace Same
p-value: 3.0e-02
Responsible for Action
p-value: 5.8e-04
Similar Situation
p-value: 3.8e-03
0 0.1 0.2 0.3 0.4 0.5 0.6 0.7 0.8 0.9 1

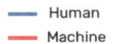

Figure 2.5 shows our data for these four scenarios. Here, we observe a gradient, ranging from a scenario in which we observe differences in judgment to one in which we don't.

In the first scenario, the one in which a flag is used to clean a bathroom (borrowed from Jonathan Haidt's work[26]), people assign strong intentionality to the human and also consider the action of the human to be more morally wrong. Unlike in most other cases, the human is liked less than the robot. This is a situation in which, compared to the robot, the human does not catch a break. Still, people prefer to replace the robot with a human more than replacing the human with a robot. But other than that, people tend to accept robots that clean bathrooms with a flag more than humans using a flag for the same purpose.

In the case of the anthem interruption, people also assign strong intentionality to the human and see the human action as slightly more morally wrong than the robot action. However, here they don't dislike the human more than the AI system.

In the wrong anthem scenario, people judge the action as unintentional. In this case, they don't see the human action as more morally wrong, and they report liking the human significantly more than the AI. This result agrees with those in previous cases describing accidents (i.e., car accidents), where the participants also tended to empathize more with human actions.

Finally, in the case of the school demolition, the actions of the human and the AI are judged equally among most dimensions. Human and robot actions are seen as equally intentional and morally wrong, and humans and robots are equally liked.

In this chapter, we got started with our empirical study of AI ethics. We compared humans and AIs making life-or-death decisions; and creating controversial ads, lewd plays, and blasphemous comedy sketches. Next, we looked at self-driving vehicles and at the desecration of national symbols. These examples showed some differences in the way in which people judge humans and machines. Yet this is only the beginning. In the next chapter, we continue our exploration by looking at cases of algorithmic bias. There is still much to learn about how humans judge machines.

Judged
by Machines

3

CHAPTER 3

In the mid-1990s, various US carriers raised an antitrust case against American Airlines and United Airlines. The complaint was that online search systems were biased against foreign and domestic carriers.[1] Their case focused on the algorithms used in ticket reservation systems. These systems prioritized flights that used the same carrier for all the legs of a trip, so an agent searching for a ticket from Louisville, Kentucky, to London, going through New York, would see flights involving no change of carrier higher on the list than flights involving two carriers. Since in the 1990s, screens showed only four to five flights, and 90 percent of all bookings came from the first screen, small differences in ranking had big financial implications.

As this airline reservation example shows, algorithms are not always fair. In fact, algorithmic bias is now a prevalent topic of discussion as it concerns computer vision systems,[2] university admission protocols,[3] natural language processing,[4] recommender systems,[5] courtroom risk assessment tools,[6] online advertisements,[7] and finance.[8] But much like human biases, algorithmic biases are not simply the result of maleficence. They can emerge from both practical and fundamental considerations.[9]

On the practical side, both people and machines learn from data that are often biased and incomplete—the data we have instead of the data we wish we had.[10]

Biased data can lead to biased learning and behavior. But even with the existence of perfect data, guaranteeing fairness may not be possible. Fairness is a concept that can be defined in multiple ways, so it is not always possible to satisfy multiple definitions of fairness simultaneously.[11]

To illustrate this fundamental limitation, consider two populations: A and B. A and B could be people identifying with different genders, or belonging to different nationalities, age groups, or ethnicities. For the purpose of this exercise, the type of difference or its source doesn't matter. What matters is that we want to achieve a fair outcome when it comes to our treatment of populations A and B.

But what constitutes a fair outcome? To keep things simple, consider two definitions.

The first definition is known as *statistical parity* or *demographic parity*. This means guaranteeing that outcomes affect equal proportions of A and B.[12] The second definition is *equality of false rejections or equality of opportunity*.[13] This means guaranteeing that the probability of being rejected if you are from population A or population B is the same.

In principle, satisfying both definitions is possible if we consider an outcome that doesn't hinge on any particular selection criterion or merit. For instance, if we pass out free concert tickets at random, we would satisfy both statistical parity and equality of false rejections. But what if the fans of the band playing in the concert were not equally distributed among both populations, A and B? In that case, distributing tickets at random would be unfair for the group that included most of these fans. Fans in this group would get fewer tickets and be more likely to be rejected.

This simple example can help us motivate more complex—and relevant—cases. Instead of free concert tickets, consider giving out loans, admitting students to college, or giving someone a promotion at work. These are all cases that not only are more delicate, but also imply some degree of selection or merit. The case of the loan is more straightforward. In principle, loans should be allocated to those who are more likely to repay them. Promotions and college admissions are trickier because they invoke the idea of merit, which may be harder to measure, even post hoc, than whether someone can repay a loan.

To illustrate how selection or merit interacts with our two notions of fairness, let populations A and B be of the same size but have a different probability of paying back a loan. To keep things simple, assume that 40 percent of the people in A would repay a loan, but only 20 percent of the people in B would.

We can achieve statistical parity by giving loans to 20 percent of the people in A and 20 percent of the people in B. This would be fair, in that the same fraction of both populations would get a loan, but would violate equality of opportunity, since we would be rejecting 20 percent of people in A who would repay their loans. But if we enforce equality of opportunity, we will end up giving more loans to people in group A, violating statistical parity.

All this goes to show that, even in simple examples, satisfying multiple definitions of fairness cannot be guaranteed. This is *not* because fairness cannot be defined, but because it allows multiple definitions. In this particular example, we used only two definitions, but we could have used many more.[14] We may include a third definition, requiring equality in false acceptances (e.g., giving loans to people who will not pay them with the same probability in both groups).

Fairness is a complex concept that accepts multiple definitions that (in most cases) cannot be satisfied simultaneously.[15] The world is unfair—not only because people and machines are biased—but because it affords multiple ways of defining a fair outcome.

Yet, not all unfairness comes from mathematical impossibilities. In fact, unfairness also comes from algorithms and the data used to design them. While in principle, these sources of unfairness could be vexing, in practice they are also sources of unfairness that potentially could be corrected.

Consider the example of word embeddings. *Word embedding* is a natural language-processing technique used to translate words into mathematical representations. It is also a popular example of algorithmic bias. This is because word embeddings can perpetuate the racial and gender stereotypes found in its training data. In a word embedding, adding the vector for the word *Queen* and that for the word *Man* gives you the word *King*. This means that these vectors satisfy semantic relationships (e.g., "a King is a male Queen"). But not all the relationships learned by word embedding are as simple and uncontroversial. Word embeddings also encode relationships, such as "Man is to computer programmer as woman is to homemaker."[16] In fact, if the text used to train the embedding contains mentions of women performing stereotypical actions, such as cooking or cleaning, the embedding will codify, maintain, and sometimes even enhance these stereotypical associations.[17]

Recent research, however, has focused not only on documenting these biases, but also on how to reduce them.[18] For instance, word embedding bias can be reduced by expanding text with sentences that counterbalance biases, or by identifying and "subtracting" the dimensions where bias manifests itself more strongly.[19]

Another example of data-driven algorithmic bias is facial recognition systems.[20] People studying the accuracy of these algorithms have found them to be less accurate at identifying darker faces, especially those of black women.[21] This has motivated the creation of data sets that are more comprehensive in terms of demographic attributes, poses, and image quality,[22] as well as the rise of auditing efforts designed to check and report on the accuracy and biases of facial recognition systems.[23]

Another discussion on algorithmic bias involves the use of pretrial "risk assessment" tools.[24] These are algorithms used to predict the probability that a defendant will reoffend (*recidivism*) or fail to appear in court.[25] Pretrial risk assessment tools have become popular in the US, but they also have been found to show biases. In 2016, investigators working for ProPublica[26] published an article based on "risk scores assigned to more than 7,000 people arrested in Broward County, Florida, in 2013 and 2014."

They used that data to "see how many were charged with new crimes over the next two years," which was "the same benchmark used by the creators of the algorithm." They found that the algorithm "was particularly likely to falsely flag black defendants as future criminals, . . . at almost twice the rate as white defendants." They also found that disparities could not be explained by prior crimes.

Unfortunately, biases are not unique to algorithms. Humans have them too. Scholars in the social sciences, for instance, have long studied the biases affecting job applications[27] by looking at the callback rates for résumés with ethnically differentiated names[28] or photographs.[29] Thus, neither humans nor machines can guarantee fairness.

Here, we compare people's reactions to cases of bias attributed to humans or machines. We present them in the context of college admissions, police enforcement, salaries, counseling, and human resources; in scenarios where humans or algorithms are the source of bias or the ones helping reduce bias. As in the previous chapter, we base our study on scenarios and measure people's reactions to them using the following questions (as appropriate):

- Were the [person/algorithm]'s actions **harmful**?
  (from "Not harmful at all" to "Extremely harmful")
- Would you **hire** this [person/algorithm] for a similar position?
  (from "Definitely not" to "Definitely yes")
- Were the [person/algorithm]'s actions **intentional**?
  (from "Not intentional at all" to "Extremely intentional")
- Do you **like** the [person/algorithm]?
  (from "Strongly dislike" to "Strongly like")
- How **morally** wrong or right were the [person/algorithm]'s actions?
  (from "Extremely wrong" to "Extremely right")
- Do you agree that the [person/algorithm] should be **promoted** to a position with more responsibilities? (from "Strongly disagree" to "Strongly agree")

- Do you agree that the [person/algorithm] should be replaced by a(n) [algorithm/person] (**replace different**)? (from "Strongly disagree" to "Strongly agree")
- Do you agree that the [person/algorithm] should be replaced by another [person/algorithm] (**replace same**)?(from "Strongly disagree" to "Strongly agree")
- Do you think the [person/algorithm] is **responsible** for the **action**)? (from "Not responsible at all" to "Extremely responsible")
- Do you think the [person/algorithm] is **responsible** for the [discriminatory/fair] **outcome**)? (from "Not responsible at all" to "Extremely responsible")
- If you were in a **similar situation** as the [person/algorithm], would you have done the same? (from "Definitely not" to "Definitely yes")

In addition to considering situations where a machine or a human either acted unfairly or corrected an unfair act, we considered variations in the ethnicity of the person being discriminated against (Hispanic, African American, or Asian). Different ethnicities are associated with different core stereotypes, so we expect different judgments in discriminatory situations.

In total, we considered a total of twenty-four possible scenarios and forty-eight conditions.* In the next section, we document the results obtained for scenarios involving human resource (HR) screenings, college admissions, salary increases, and policing.

The four groups of scenarios are listed next.

---

* Certainly, we would have liked to consider more conditions, such as additional ethnicities and nonbinary gender identities, but that would have increased the number of scenarios and independent groups that we had to recruit to an unwieldy number. We leave the exercise of extending this analysis to more conditions for the future.

Human Resource Screenings

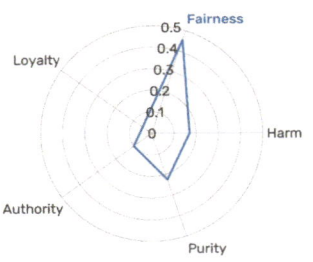

A company replaces their HR manager with a new [manager/algorithm] tasked with screening candidates for job interviews.

Unfair treatment

An audit reveals that the new [manager/algorithm] never selects [Hispanic/African American/Asian] candidates even when they have the same qualifications as other candidates.

Fair treatment

An audit reveals that the new [manager/algorithm] produces a fairer process for [Hispanic/African American/Asian] candidates, who were discriminated against by the previous system.

College Admissions

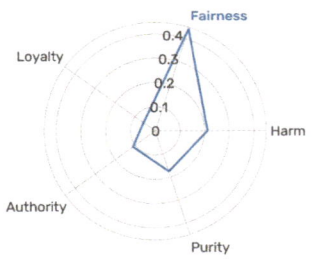

To improve their admissions process, a university hires a new [recruiter/algorithm] to evaluate the grades, test scores, and recommendation letters of applicants.

S25 **S26** **S27**

Unfair treatment
An audit reveals that the new [recruiter/algorithm] is biased against [Hispanic/African American/Asian] applicants.

S28 **S29** **S30**

Fair treatment
An audit reveals that the new [recruiter/algorithm] is fairer to [Hispanic/African American/Asian] applicants, who were discriminated against by the previous system.

Salary Increases

  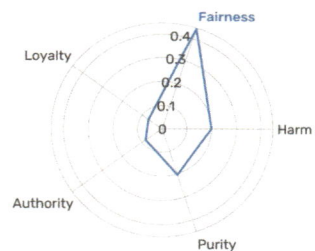

A financial company hires a new [manager/algorithm] to decide the yearly salary increases of its employees.

S31 **S32** **S33**

Unfair treatment
An audit reveals that the new [manager/algorithm] consistently gives lower raises to [Hispanic/African American/Asian] employees, even when they are equal to other employees.

S34 **S35** **S36**

Fair treatment:
An audit reveals that the new [manager/algorithm] is fairer to [Hispanic/African American/Asian] employees, who were being discriminated against by the previous process.

Policing

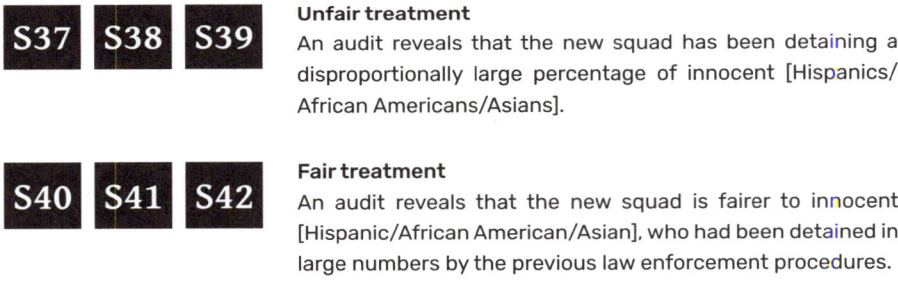

The police commissioner of a major city deploys a new squad of [police officers/police robots] in a high-crime neighborhood.

S37 S38 S39

Unfair treatment
An audit reveals that the new squad has been detaining a disproportionally large percentage of innocent [Hispanics/ African Americans/Asians].

S40 S41 S42

Fair treatment
An audit reveals that the new squad is fairer to innocent [Hispanic/African American/Asian], who had been detained in large numbers by the previous law enforcement procedures.

Figure 3.1 shows people's reactions to the scenarios in which discrimination was observed. In all cases, humans are seen as more intentional, and also as more responsible for actions and outcomes. But beyond these obvious effects, we do observe some interesting, albeit relatively weak, patterns.

First, we find that—unlike most previous cases—moral judgments are not favorable to humans. In fact, we find that human actions are judged worse than machine actions (i.e., less moral), and are seen as more harmful in several scenarios, such as the college admissions and salary scenarios for African Americans and Hispanics. This provides additional evidence supporting the idea that reactions to machine actions are not simply the result of a generalized bias against machines since these biases change with a scenario's context and moral dimensions.

We also find small but interesting differences among the various ethnic groups depending on the particular scenario. The college admissions scenario elicits the strongest differences in judgment, especially for African Americans and Hispanics (who suffer more discrimination than Asians in contexts related to intelectual traits because of differences in stereotypes). Here, human actions are judged as relatively less moral and more harmful than the actions of machines. We also find that biases against African Americans and Hispanics result in slightly stronger differences in judgment between humans and machines compared to Asians. This suggests that differences in judgment between human and machine actions are slightly modulated by the ethnic group of the victim and the situation described in the scenario (e.g., college admissions). These differences aligns with our expectations for a US sample.

Moreover, people also think that the human should be replaced with another person. What is paradoxical, however, is that even though humans are seen as more intentional and more responsible than machines, people still prefer not to replace them with machines (as has been the case in all previous scenarios), adding further evidence in support of the idea of algorithm aversion.[30]

Figure 3.2 shows people's reactions to scenarios in which discrimination was corrected. In general, we find a tendency for people to be more willing to promote humans, meaning that humans may receive more credit when they are involved in actions that correct unfair treatment. For the most part, however, we don't find strong differences in judgment, except for the policing scenario, where the actions of humans are judged as much better than those of machines across several dimensions.

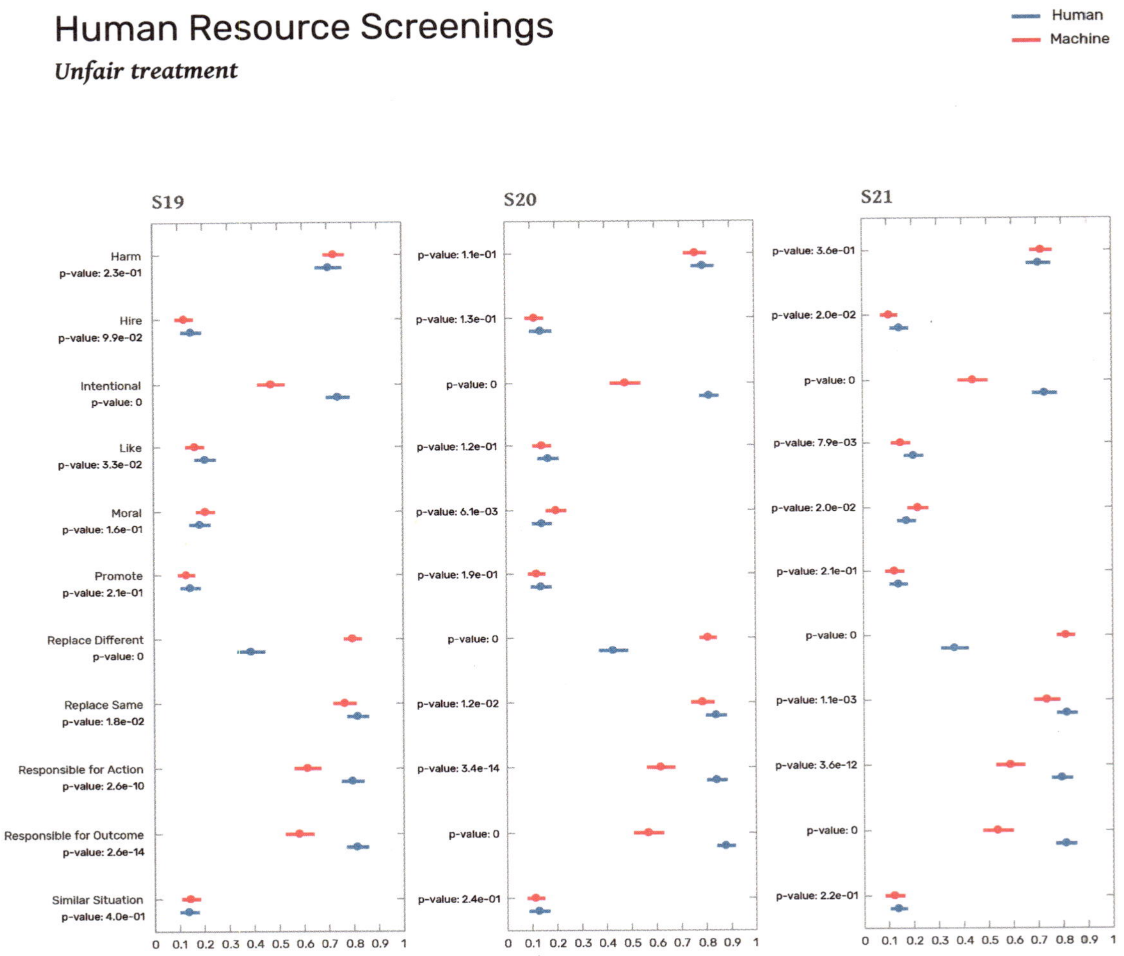

Human Resource Screenings

Unfair treatment

— Human
— Machine

Figure 3.1

Participant reactions to four discrimination scenarios:
Human Resource Screenings (S19,S20,S21), College Admissions (S25,S26,S27),
Salary Increases (S31,S32,S33) and Policing (S37,S38,S39).

College Admissions
Unfair treatment

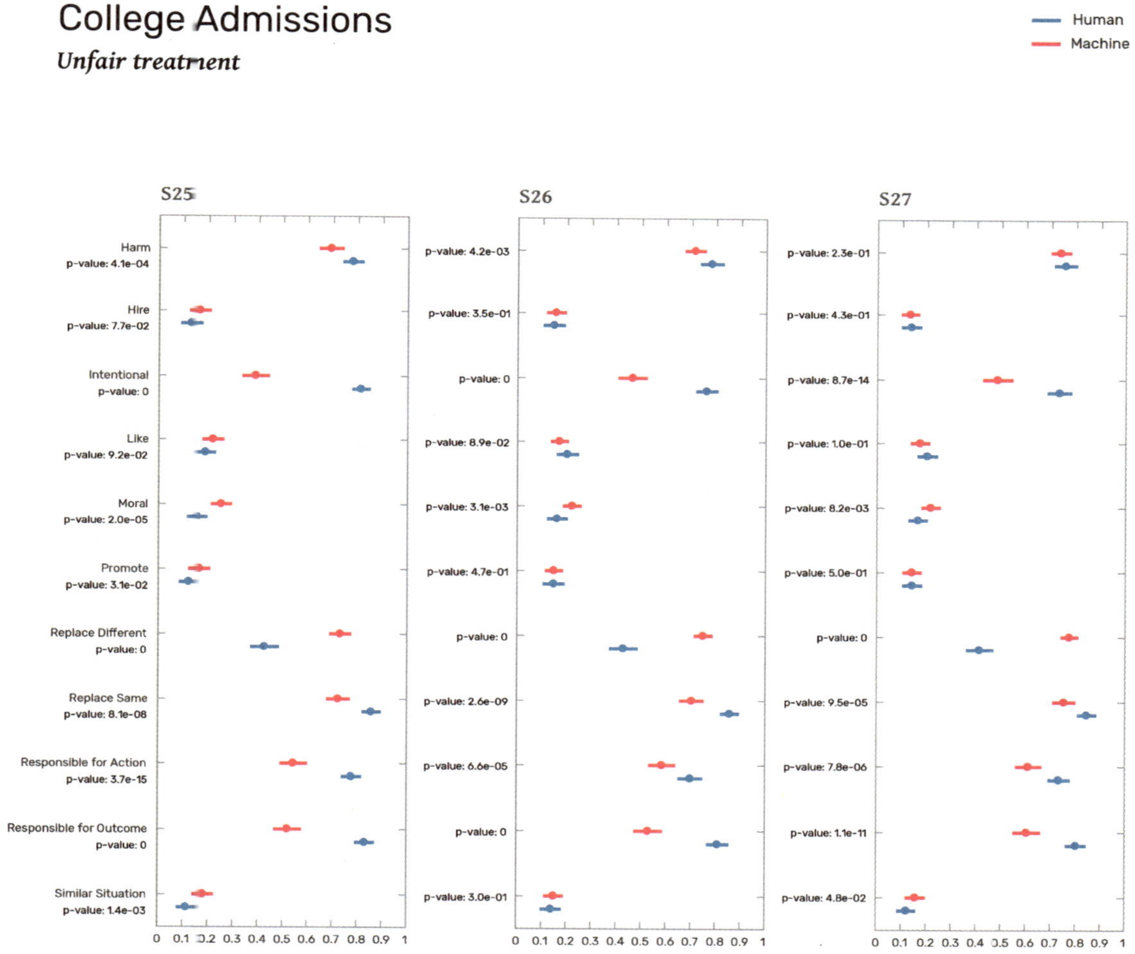

Salary Increases
Unfair treatment

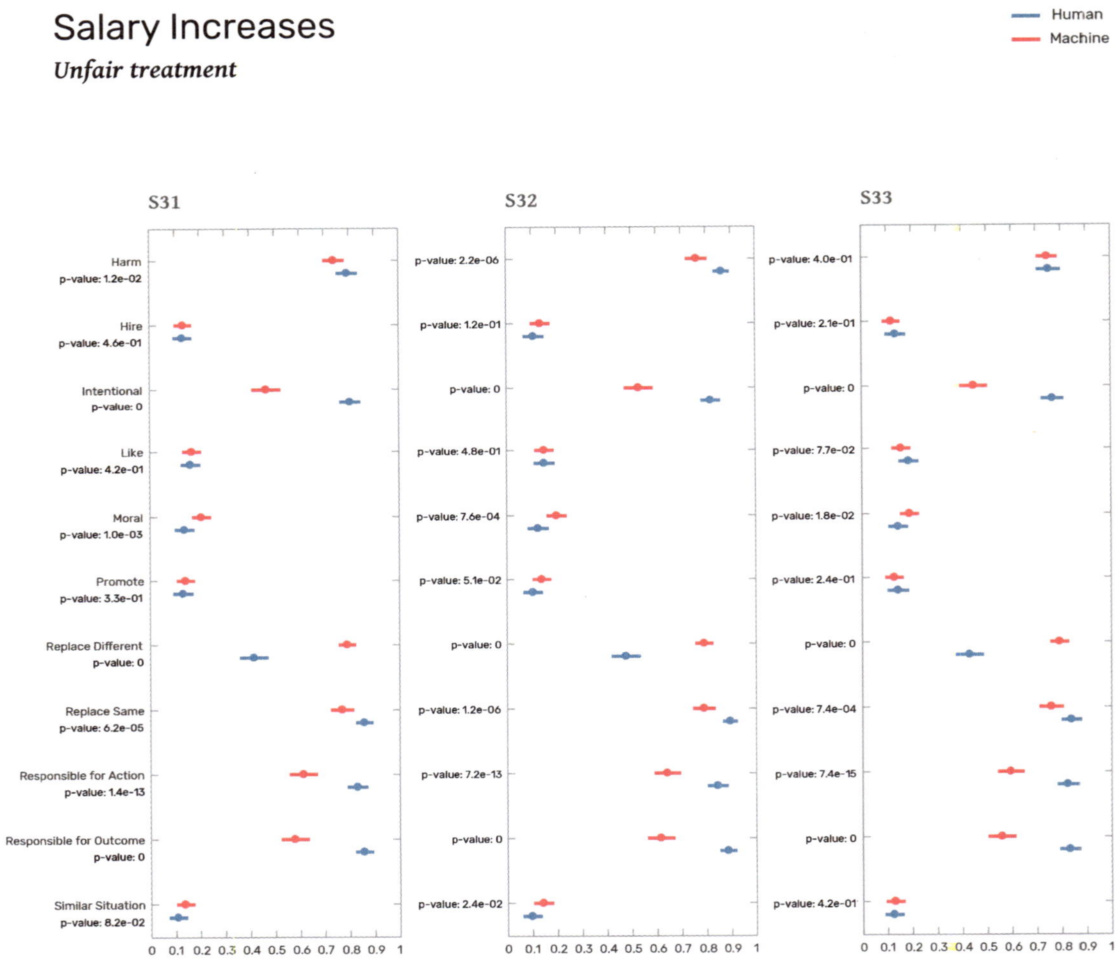

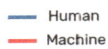

Human
Machine
S31
Harm
p-value: 1.2e-02
Hire
p-value: 4.6e-01
Intentional
p-value: 0
Like
p-value: 4.2e-01
Moral
p-value: 1.0e-03
Promote
p-value: 3.3e-01
Replace Different
p-value: 0
Replace Same
p-value: 6.2e-05
Responsible for Action
p-value: 1.4e-13
Responsible for Outcome
p-value: 0
Similar Situation
p-value: 8.2e-02
0 0.1 0.2 0.3 0.4 0.5 0.6 0.7 0.8 0.9 1
S32
p-value: 2.2e-06
p-value: 1.2e-01
p-value: 0
p-value: 4.8e-01
p-value: 7.6e-04
p-value: 5.1e-02
p-value: 0
p-value: 1.2e-06
p-value: 7.2e-13
p-value: 0
p-value: 2.4e-02
0 0.1 0.2 0.3 0.4 0.5 0.6 0.7 0.8 0.9 1
S33
p-value: 4.0e-01
p-value: 2.1e-01
p-value: 0
p-value: 7.7e-02
p-value: 1.8e-02
p-value: 2.4e-01
p-value: 0
p-value: 7.4e-04
p-value: 7.4e-15
p-value: 0
p-value: 4.2e-01
0 0.1 0.2 0.3 0.4 0.5 0.6 0.7 0.8 0.9 1

Policing

Unfair treatment

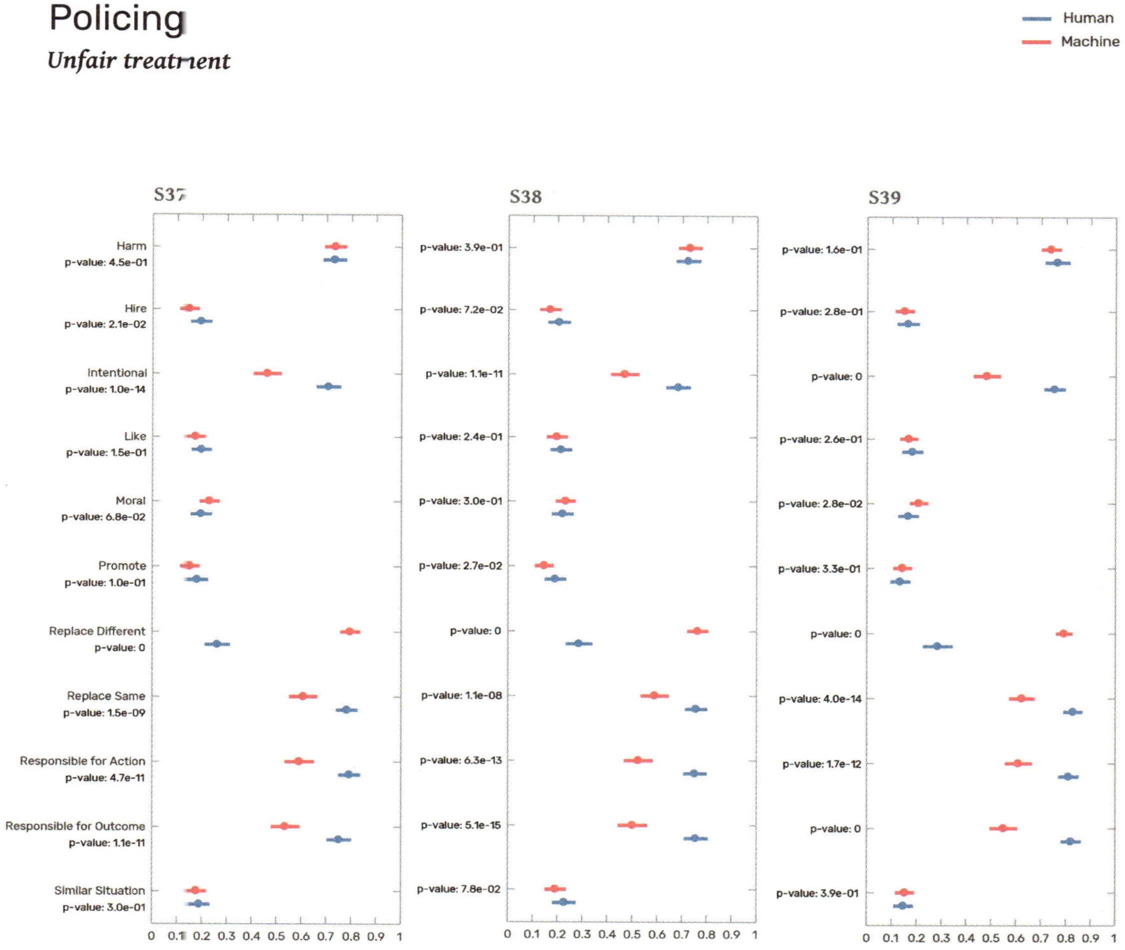

Human
Machine

S37

Harm	p-value: 4.5e-01	
Hire	p-value: 2.1e-02	
Intentional	p-value: 1.0e-14	
Like	p-value: 1.5e-01	
Moral	p-value: 6.8e-02	
Promote	p-value: 1.0e-01	
Replace Different	p-value: 0	
Replace Same	p-value: 1.5e-09	
Responsible for Action	p-value: 4.7e-11	
Responsible for Outcome	p-value: 1.1e-11	
Similar Situation	p-value: 3.0e-01	

0 0.1 0.2 0.3 0.4 0.5 0.6 0.7 0.8 0.9 1

S38

p-value: 3.9e-01
p-value: 7.2e-02
p-value: 1.1e-11
p-value: 2.4e-01
p-value: 3.0e-01
p-value: 2.7e-02
p-value: 0
p-value: 1.1e-08
p-value: 6.3e-13
p-value: 5.1e-15
p-value: 7.8e-02

0 0.1 0.2 0.3 0.4 0.5 0.6 0.7 0.8 0.9 1

S39

p-value: 1.6e-01
p-value: 2.8e-01
p-value: 0
p-value: 2.6e-01
p-value: 2.8e-02
p-value: 3.3e-01
p-value: 0
p-value: 4.0e-14
p-value: 1.7e-12
p-value: 0
p-value: 3.9e-01

0 0.1 0.2 0.3 0.4 0.5 0.6 0.7 0.8 0.9 1

Human Resource Screenings

Fair treatment

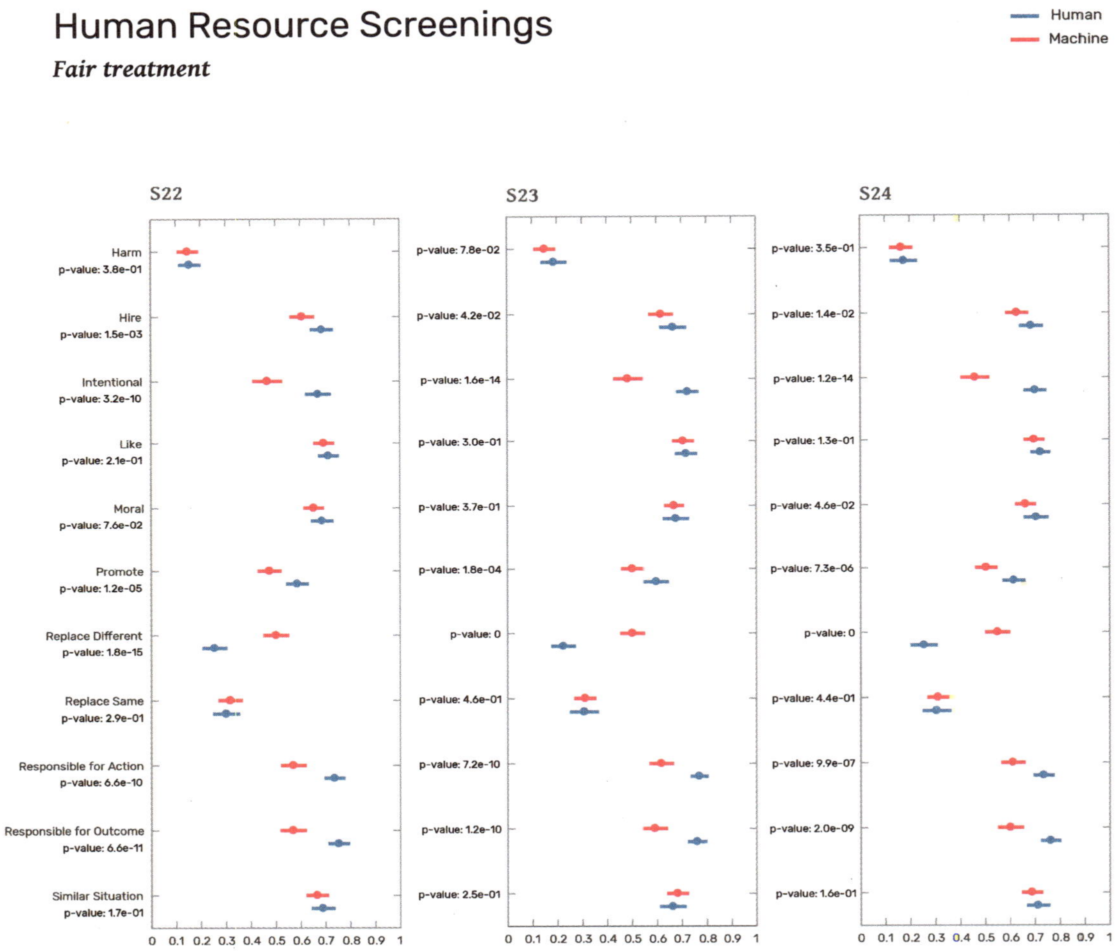

	S22	S23	S24
Harm	p-value: 3.8e-01	p-value: 7.8e-02	p-value: 3.5e-01
Hire	p-value: 1.5e-03	p-value: 4.2e-02	p-value: 1.4e-02
Intentional	p-value: 3.2e-10	p-value: 1.6e-14	p-value: 1.2e-14
Like	p-value: 2.1e-01	p-value: 3.0e-01	p-value: 1.3e-01
Moral	p-value: 7.6e-02	p-value: 3.7e-01	p-value: 4.6e-02
Promote	p-value: 1.2e-05	p-value: 1.8e-04	p-value: 7.3e-06
Replace Different	p-value: 1.8e-15	p-value: 0	p-value: 0
Replace Same	p-value: 2.9e-01	p-value: 4.6e-01	p-value: 4.4e-01
Responsible for Action	p-value: 6.6e-10	p-value: 7.2e-10	p-value: 9.9e-07
Responsible for Outcome	p-value: 6.6e-11	p-value: 1.2e-10	p-value: 2.0e-09
Similar Situation	p-value: 1.7e-01	p-value: 2.5e-01	p-value: 1.6e-01

Figure 3.2

Participant reactions to four corrected discrimination scenarios:
Human Resource Screenings (S22,S23,S24), College Admissions (S28,S29,S30),
Salary Increases (S34,S35,S36) and Policing (S40,S41,S42).

College Admissions
Fair treatment

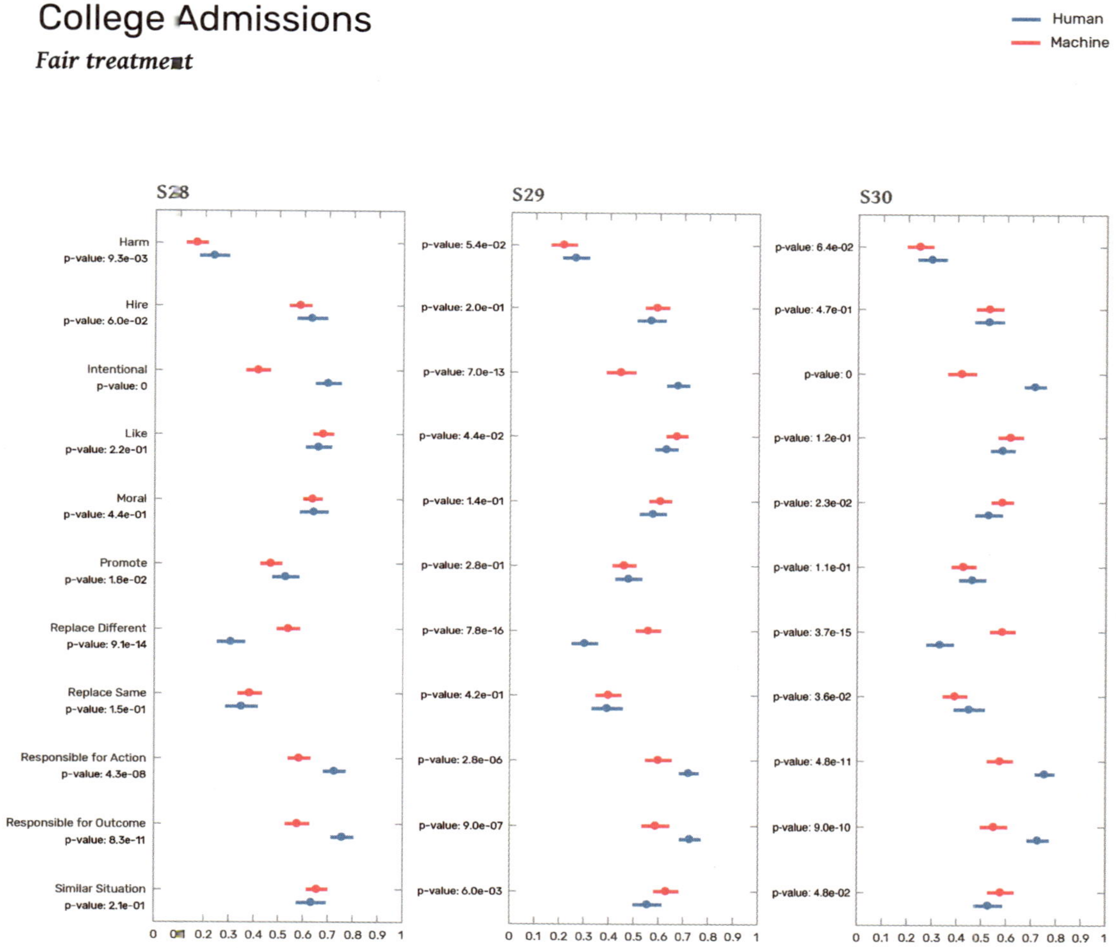

—— Human
—— Machine

Salary Increases

Fair treatment

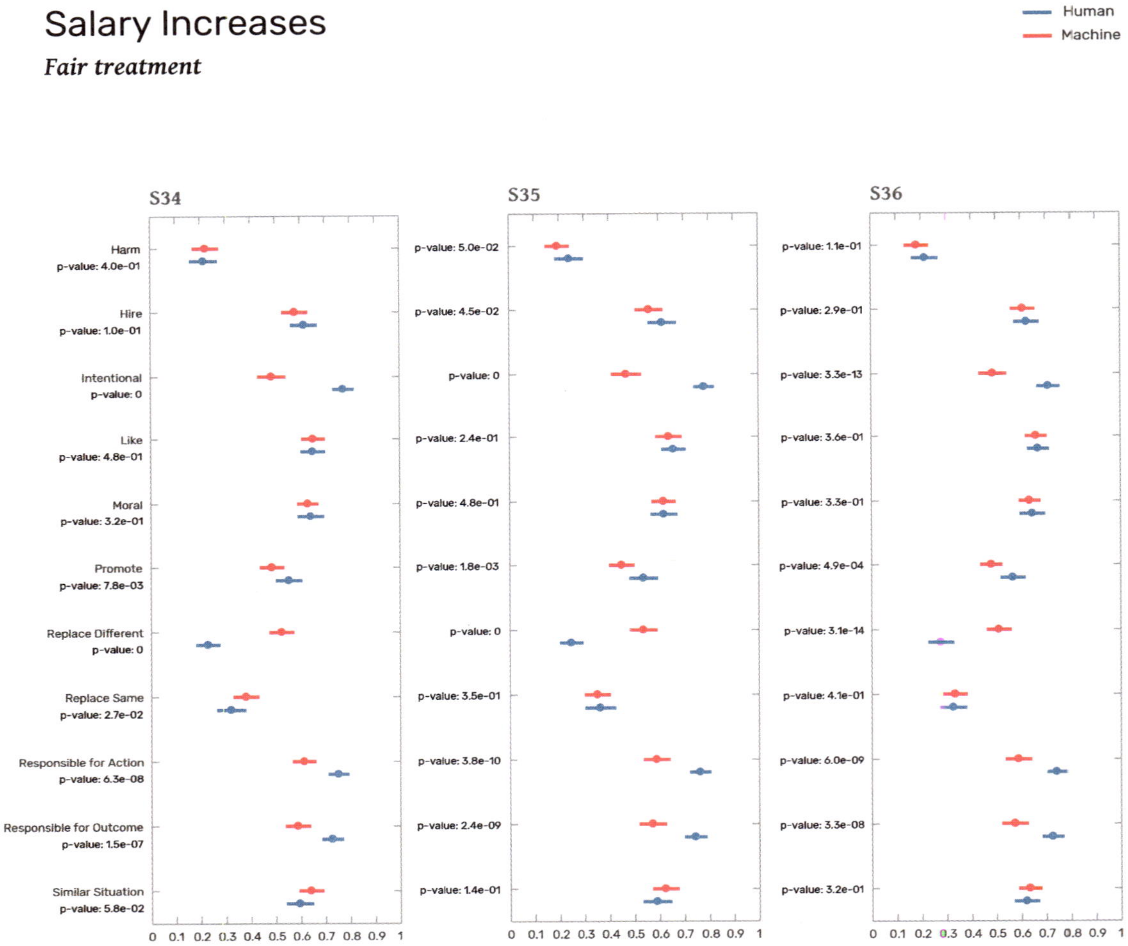

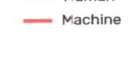

Human
Machine
S34
Harm
p-value: 4.0e-01
Hire
p-value: 1.0e-01
Intentional
p-value: 0
Like
p-value: 4.8e-01
Moral
p-value: 3.2e-01
Promote
p-value: 7.8e-03
Replace Different
p-value: 0
Replace Same
p-value: 2.7e-02
Responsible for Action
p-value: 6.3e-08
Responsible for Outcome
p-value: 1.5e-07
Similar Situation
p-value: 5.8e-02
0 0.1 0.2 0.3 0.4 0.5 0.6 0.7 0.8 0.9 1
S35
p-value: 5.0e-02
p-value: 4.5e-02
p-value: 0
p-value: 2.4e-01
p-value: 4.8e-01
p-value: 1.8e-03
p-value: 0
p-value: 3.5e-01
p-value: 3.8e-10
p-value: 2.4e-09
p-value: 1.4e-01
0 0.1 0.2 0.3 0.4 0.5 0.6 0.7 0.8 0.9 1
S36
p-value: 1.1e-01
p-value: 2.9e-01
p-value: 3.3e-13
p-value: 3.6e-01
p-value: 3.3e-01
p-value: 4.9e-04
p-value: 3.1e-14
p-value: 4.1e-01
p-value: 6.0e-09
p-value: 3.3e-08
p-value: 3.2e-01
0 0.1 0.2 0.3 0.4 0.5 0.6 0.7 0.8 0.9 1

Policing
Fair treatment

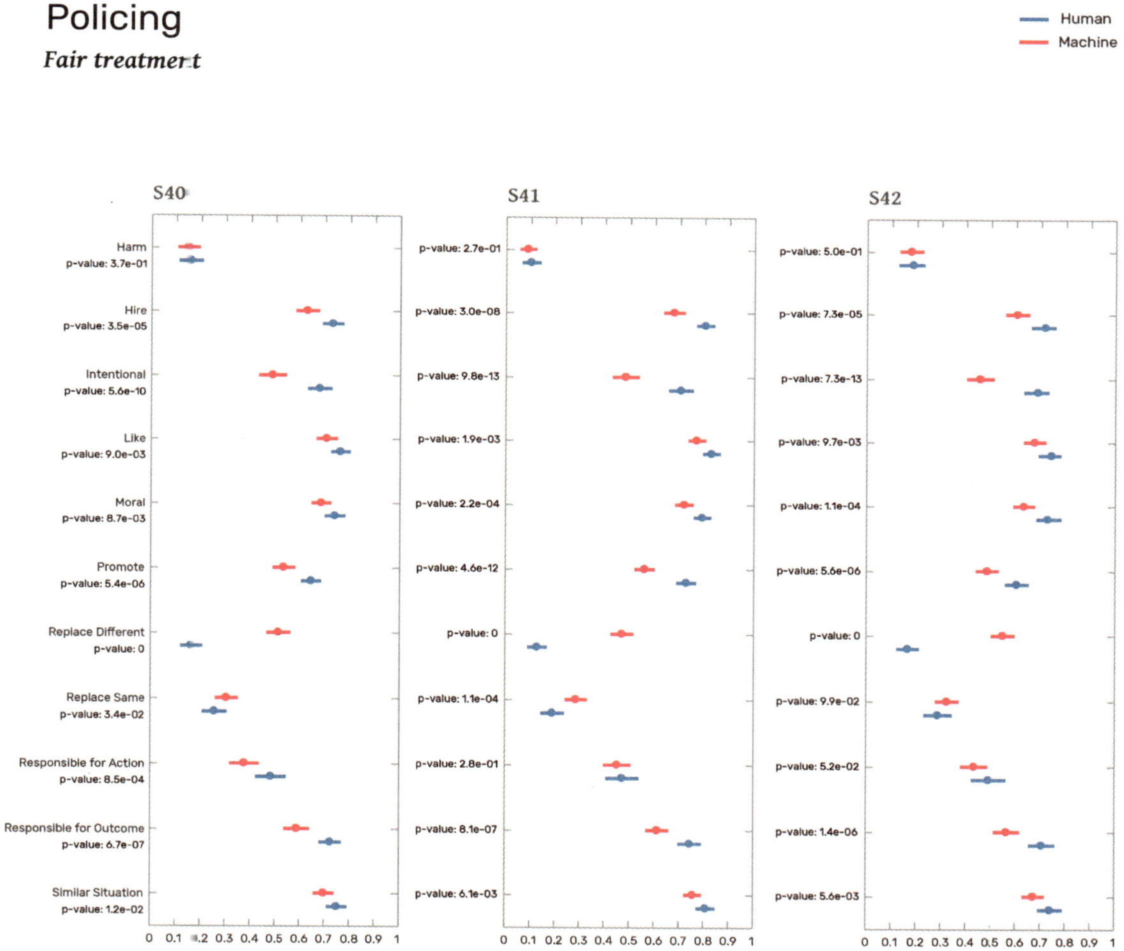

— Human
— Machine

S40

Harm	p-value: 3.7e-01
Hire	p-value: 3.5e-05
Intentional	p-value: 5.6e-10
Like	p-value: 9.0e-03
Moral	p-value: 8.7e-03
Promote	p-value: 5.4e-06
Replace Different	p-value: 0
Replace Same	p-value: 3.4e-02
Responsible for Action	p-value: 8.5e-04
Responsible for Outcome	p-value: 6.7e-07
Similar Situation	p-value: 1.2e-02

0 0.1 0.2 0.3 0.4 0.5 0.6 0.7 0.8 0.9 1

S41

p-value: 2.7e-01
p-value: 3.0e-08
p-value: 9.8e-13
p-value: 1.9e-03
p-value: 2.2e-04
p-value: 4.6e-12
p-value: 0
p-value: 1.1e-04
p-value: 2.8e-01
p-value: 8.1e-07
p-value: 6.1e-03

0 0.1 0.2 0.3 0.4 0.5 0.6 0.7 0.8 0.9 1

S42

p-value: 5.0e-01
p-value: 7.3e-05
p-value: 7.3e-13
p-value: 9.7e-03
p-value: 1.1e-04
p-value: 5.6e-06
p-value: 0
p-value: 9.9e-02
p-value: 5.2e-02
p-value: 1.4e-06
p-value: 5.6e-03

0 0.1 0.2 0.3 0.4 0.5 0.6 0.7 0.8 0.9 1

While biases can be problematic, the psychologists who have long studied them would be hard pressed to classify them as simple cognitive flaws. Instead, biases exist as rules of thumb or heuristics that evolved to make fast decisions in environments with limited information.[31]

An example of these heuristics is the idea that people may perceive groups using two dominant characteristics: warmth and competence.[32] This model predicts that groups high in warmth and low in competence (e.g., disabled people, babies, and the elderly) elicit sympathy, whereas groups low in warmth and high in competence elicit envy or jealousy.

Heuristics and stereotypes are certainly incorrect ways to judge individuals. Humans can have overgeneralized beliefs regarding members of a social group (stereotypes) and exhibit biased attitudes toward those groups (prejudice). Prejudice may then lead to unfair treatment or discrimination. But because heuristics work as a way to facilitate decision-making in information-deprived environments (or environments with excess information),[33] it is not surprising that we find them in both humans and machines.

Similar to humans, cognitive machines are inferential and base their inferences on abstract forms of categorization (which is called *stereotyping* in humans). In order to make predictions, machines often group and classify data using explicit and abstract features. Consider the idea of a principal component—a vector that accounts for most of the variance in a data set. Principal components are a common tool in machine learning, and they are similar to the idea of a stereotype, like classifying people using the vectors of warmth and competence. Unlike warmth and competence, however, principal components usually involve abstract features that are derived directly from data and can be difficult to interpret. This adds obscurity to algorithms and has led some people to advocate for increased transparency and interpretability as ways to mitigate algorithmic bias.

In fact, the use of explicit versus abstract features has been at the core of a nuanced discussion on the bias and fairness of algorithms lately. To avoid biases based on gender or race, scholars have proposed a variety of methods, from simply removing explicit demographic characteristics from a data set to predicting outcomes using only variables that are orthogonal to demographic characteristics. Yet recent research has shown that methods that tend to circumvent the possibility of bias may actually backfire because reaching fair outcomes is better served by using the most accurate predictors, even if these include explicit demographic information.[34]

In a recent paper on algorithmic fairness, Jon Kleinberg, Jens Ludwig, Sendhil Mullainathan, and Ashesh Rambachan develop this idea by comparing an *efficient* and an *equitable*.[35] These planners were tasked with admitting college applicants. The efficient planner was interested only in maximizing performance, measured by the grade point average (GPA) of the students admitted to college, while the equitable planner was interested in both performance and the racial composition of the admitted class.

To illustrate this idea, the scholars compared three methods: admissions that were blind to demographic variables (e.g., race was removed from the sample); admissions that included variables that were orthogonalized with respect to racial variables; and admissions that used racial variables explicitly. They report that the most equitable and efficient outcomes were reached using the model that explicitly included demographic variables.

To understand this distinction, consider students from two races: P (privileged) and U (underprivileged), who are applying to college. Because of their privileges, students in race P score higher in many of the variables that are predictive of future academic success, such as standardized test scores. Should we blind algorithms to race, then? Or is there a better solution?

Imagine a student from race U that obtains the same score as a student from race P on a standardized test. The student from race U was able to reach the same outcome as the student from race P in the absence of P's privileges. Yet a model lacking an explicit racial variable will be unable to adjust for the lack of privilege affecting the scores of students from race U. A model that is blind to race will rate both students equally, and hence hurt the less privileged student. Instead, what the proposed theory suggests[36] is to use the most accurate possible model (including racial variables, when relevant), and then setting different thresholds to achieve the desired level of equity (using, for instance, some of the definitions of fairness introduced earlier in this chapter).

This example illustrates the importance of separating the goals of equity and predictive accuracy. Even though it may be tempting to modify data to eliminate any trace of demographic characteristics, the best way to achieve efficient and equitable outcomes may be to treat prediction and equity as two separate parts of the same problem.

In the US, discriminatory treatment is not only frowned upon, but also illegal. Title VII of the 1964 Civil Rights Act[37] is "a federal law that prohibits employers from discriminating against employees on the basis of sex, race, color, national origin, and religion."[†] The Supreme Court affirmed Title VII unanimously in 1971 in *Griggs v. Duke Power Company*, a class action suit claiming that Duke's policies discriminated against African American employees.[38] The court ruled that, independent of intent, discriminatory outcomes for protected classes violated Title VII.[39]

In our data, we find important differences in the level of intent and responsibility assigned to discriminatory actions performed by humans and machines. However, in agreement with the *Griggs* decision, we find only small differences in moral judgment,

---

[†] It generally applies to employers with fifteen or more employees, including federal, state, and local governments.

suggesting that—in unfair cases—it is the outcome rather than the intention that is judged.

The removal of intent from the legal judgment of bias has important implications for those working on the fairness of algorithms. It means that, even in the absence of intent, those creating the algorithms may be liable for biased outcomes.

This outcome-based approach to policing discrimination is opening a new market for an algorithm certification industry and discipline:[40] a community focused on auditing the bias of algorithms and certifying them when they are not biased.

In the next chapter, we shift our gaze away from algorithmic bias and focus on another uncomfortable aspect of our digital reality: privacy. This will help us expand our understanding to another dimension of the way in which humans judge machines.

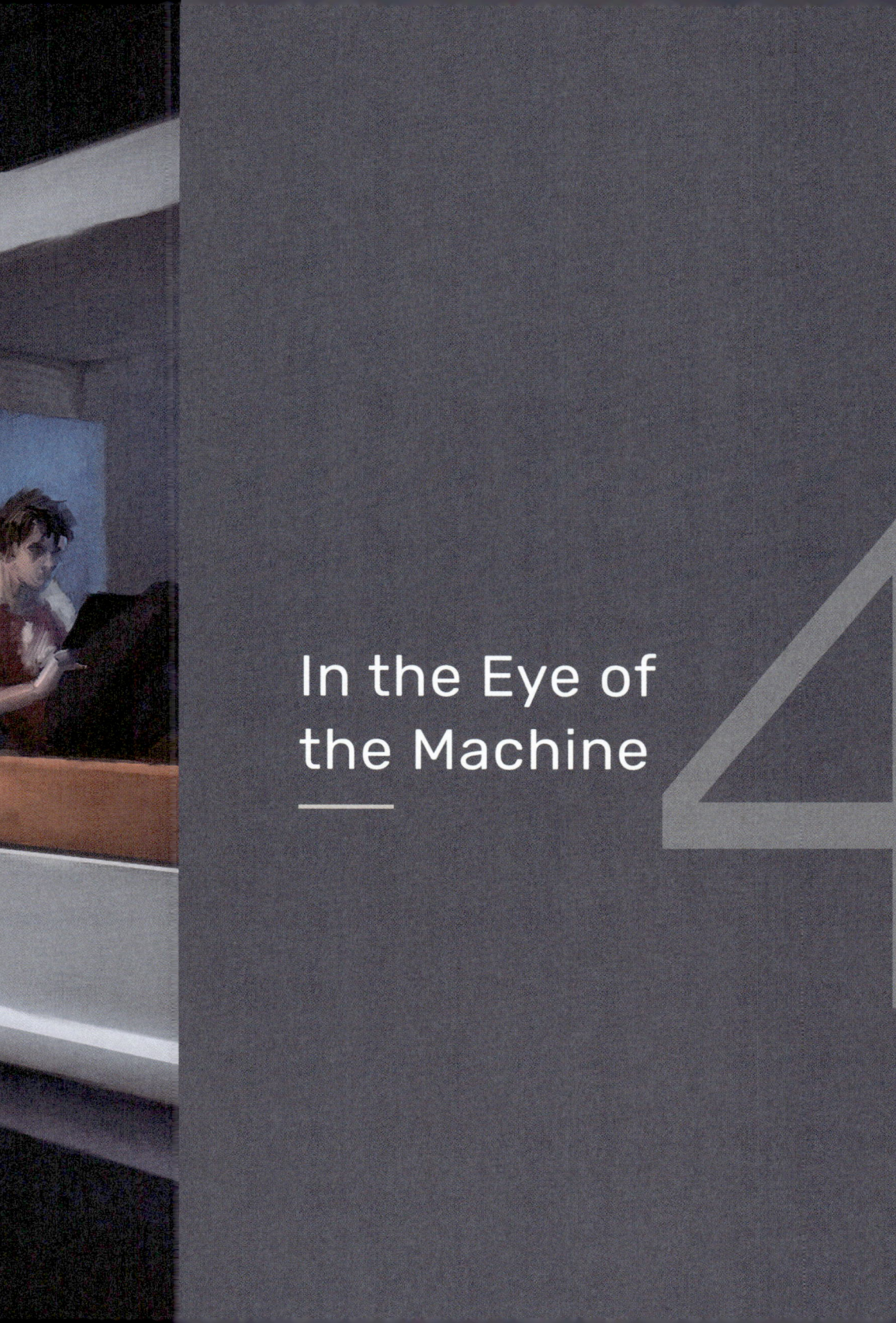

In the Eye of
the Machine
___

4

CHAPTER 4

Have your ever feared that someone is watching you?

In October 2019, the "Japanese hotel chain, HIS Group . . . apologized for ignoring warnings that its in-room robots were hackable."[1] The hack was revealed on Twitter[2] by "a security researcher [who] warned [the hotel that] the bed-bots [were] easily accessible." This vulnerability allowed "individuals to remotely view video footage from the devices [using a] streaming app." This meant that the in-room robots could potentially have been used to make a candid livestream of a customer's hotel stay.[3]

But camera bots are only a small example of the growing interface between technology and privacy.[4] On one hand, we have computer vision systems, like those embedded in the glasses of Chinese police forces[5] or in public cameras.[6] On the other hand, we have digital records, like those collected by hospitals, insurance providers, search engines, social media platforms, online retailers, mobile phone companies, and voice assistants, such as Alexa or Siri.

What both computer vision systems and data-driven platforms have in common is that they often use the data they collect to train machine-learning algorithms.

This tells us that when it comes to privacy, we need to worry about both the data that can be revealed and the information that can be revealed by models built on this data.

When it comes to data privacy, people are concerned about the possibility of identifying individuals, or gathering sensitive information about groups. When it comes to models, people are concerned about someone learning personal information by interrogating a model. This includes knowing whether a person was part of the data set used to train the model. After all, simply being part of a data set could involve sensitive information (e.g., knowing the mere fact that a person is part of a data set of cancer patients or intelligence agents would reveal sensitive information about that person even if they cannot be pinpointed in the data set).

Reidentification risks are real.[7] A famous story from 1997 involves Latanya Sweeney, now a professor at Harvard but at that time a graduate student at the Massachusetts Institute of Technology (MIT). Sweeney was able to reidentify the medical records of Massachusetts governor William Weld by using publicly available information in an anonymized data set released by the state's Group Insurance Commission.[8] Such reidentification is possible when data entries are characterized by quasi-identifiers, such as ZIP code, sex, and birthdate, that combine to form unique identifiers. Yet quasi-identifiers can also emerge spontaneously from data that has been stripped of any individual characteristics. Consider mobile phone traces. In 2013, a study using mobile phone records found that[9] "in a dataset where the location of an individual is specified hourly, and with a spatial resolution equal to . . . the carrier's antennas, four spatio-temporal points are enough to uniquely identify 95% of the individuals."

During the last few decades, scholars have proposed several methods to protect privacy. One of these is the concept of k-anonymity, proposed by Sweeney herself.[10] This is the idea that any combination of quasi-identifiers should match at least k individuals. But k-anonymity has a few problems,[11] since identifying a person within a group of *k* others can also reveal sensitive information. For instance, in a medical record, we may identify someone within a group of three people who have been diagnosed with HIV, colon cancer, and lupus. Knowing that a person has any of these three conditions constitutes sensitive information. In a real-world example, a fitness company called Strava[12] released an aggregate data visualization of the jogging routes of users of its fitness app, inadvertently releasing information about the location of military bases in Afghanistan. In addition, *k*-anonymity cannot be guaranteed when data sets are combined.[13] For instance, a person who has been to two hospitals that release *k*-anonymous data could be identified by combining these data sets.

These limitations have inspired people to think more creatively about privacy. After all, when we think of attacks on data sets protected by *k*-anonymity, we are thinking about the inferences that a person can make from data. Hence, it is reasonable to think of privacy in terms not only of data, but also of the inferences that we can make from models built on such data. This move from data to models has motivated another approach to data protection, known as *differential privacy*.[14]

In simple terms, differential privacy guarantees that an outside observer cannot know whether a person is part of a data set. This is guaranteed by ensuring that the outcome of any calculations done using the data set does not change—or does not change enough—whether a person is part of a data set or not. As Michael Kearns and Aaron Roth explain in their book *The Ethical Algorithm*:[15] "Suppose some outside observer is trying to guess whether a particular person—say, Rebecca—is in the dataset of interest." If the observer is shown the output of a computation with or without Rebecca's data, "he will not be able to guess which output was shown more accurately than random guessing."

One simple algorithm that can be used to implement differential privacy is *randomized response*.[16] This algorithm, dating back to the 1960s, can be easily explained using an example. Imagine that you want to run a survey to determine how many students in a school use drugs. In principle, drug users may not want to respond to a direct question like "Do you use drugs?" because that information could be used against them. Self-censoring would be true even if you promised to keep the data private because the information could be stolen or subpoenaed by law enforcement.[17]

Randomized response offers a solution to this problem by asking people instead to flip a coin (and keep the result confidential), and give true answers only if the coin lands on heads. If it lands on tails, people should flip the coin again and answer "yes" if the coin landed on heads and "no" if it landed on tails. This method helps reveal information, but also gives respondents plausible deniability (since the coin-flip results were never recorded).

Unfortunately, randomized response is far from bulletproof. It works well if you ask each person to respond only once. But if you ask people to respond multiple times, you become more certain about their true state with each response.[18] Also, even though randomized response works well with helping people reveal sensitive information,[19] it doesn't guarantee trust. In fact, people's trust in the method depends on their ability to understand the procedure.[20] That's why, in recent years, we have seen the rise of more sophisticated privacy-preserving algorithms, such as Rappor, PATE, Federated Learning, and Split Learning.[21]

These methods show some of the work that has gone into understanding and protecting privacy in our digital world. But how do these privacy concerns change when the agent behind the data collection efforts is a machine? The reminder of this chapter will be dedicated to exploring people's reactions to scenarios in which people are observed by humans or machines.

Consider the following six scenarios:

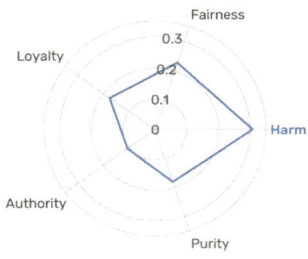

S43 A school is looking to improve the attendance and attention of its students. The school board decides to hire [people/a facial recognition system] to observe students during classes and track the attendance, emotions, and attention of each student.

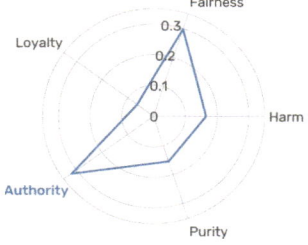

S44 In a city, students are given an ID card that allows them to ride public transportation free of charge. City workers discover that many students are cheating the system by sharing their cards with nonstudent family members. The local government decides to start checking the identity of each rider. To check if the rider's face matches the photo on the ID, an [inspector/facial recognition system] is placed at every access point. The [inspector/facial recognition system] remembers the face of every person checked.

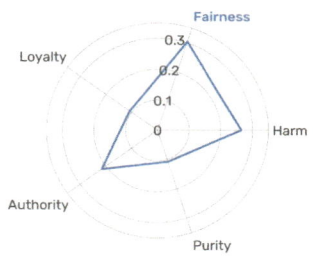

S45

A mall is looking to reduce shoplifting. To improve security, the mall decides to employ a [team of security guards/facial recognition system] to screen everyone who enters or exits the mall. The [team of security guards/facial recognition system] remembers most of the faces screened.

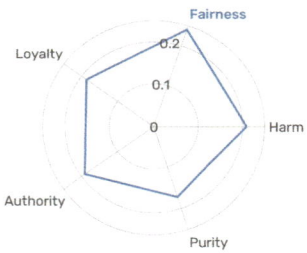

S46

A hotel is looking to build a new poolside bar. The hotel decides to equip the bar with [workers/robots] trained to recognize the face of each guest to keep track of their bills. The [workers/robots] remember everyone they see next to the pool.

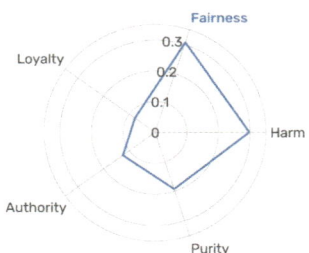

S47

An airport management team is seeking to increase security. To do so, the management team decides to equip the airport with [security officers/facial recognition cameras] that will register the face of everyone who enters the airport and track their movements.

We asked people to react to these scenarios by answering the following five questions:

- How **comfortable** are you with this system? (from "Extremely uncomfortable" to "Extremely comfortable")

- How **morally wrong or right** is this system? (from "Extremely wrong" to "Extremely right")

- Do you think this solution **violates** people's privacy? (from" Strongly disagree" to "Strongly agree")

- Would you **recommend** this system to a friend? (from "Would surely not recommend" to "Would surely recommend")

- Would you **use** this system? (from "Would surely not use" to "Would surely use")

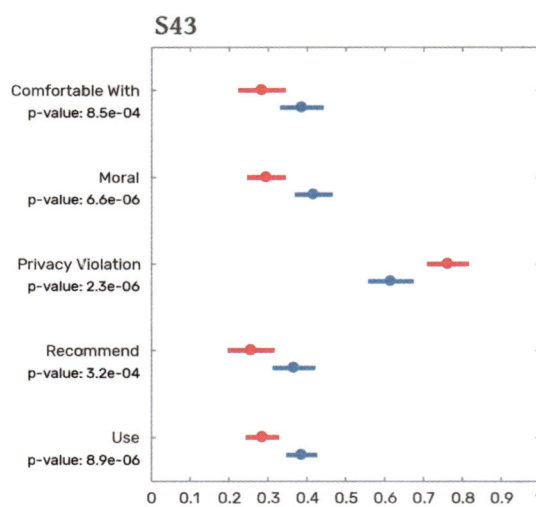

Figure 4.1

Participant reactions to five privacy scenarios: (S43,S44,S45,S46,S47).

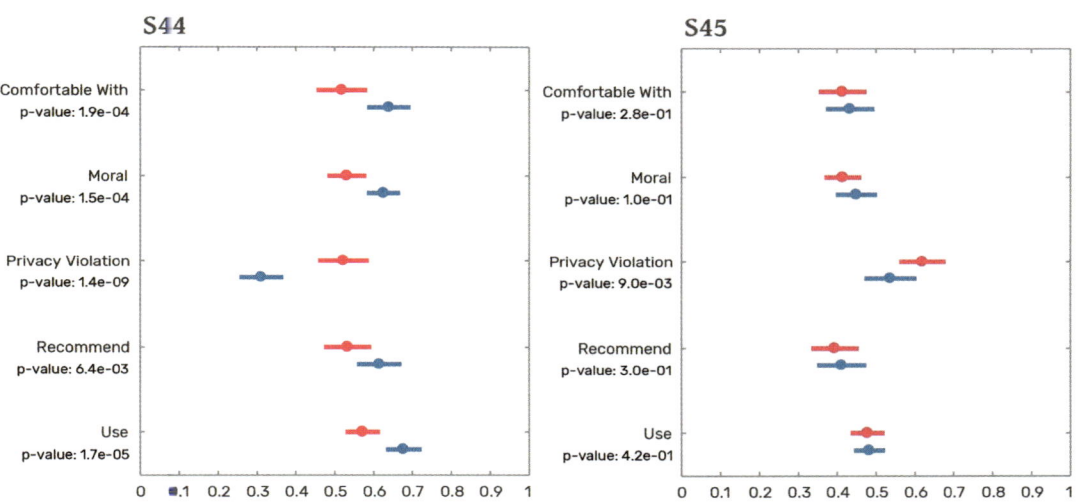

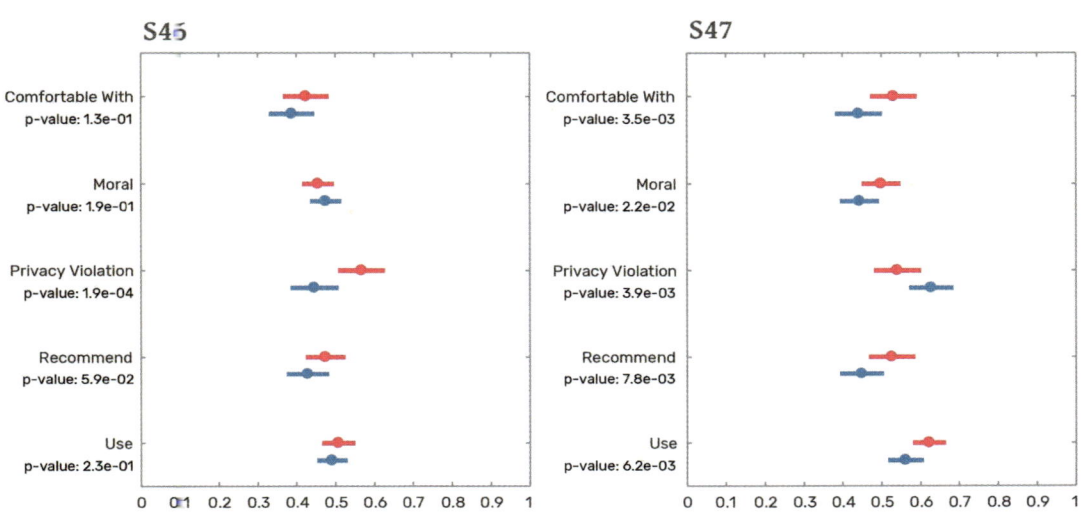

Figure 4.1 presents the result of this exercise. It reveals that people's preference for privacy depends strongly on the circumstances of each scenario. Figures 4.1a and 4.1b show the results for the school monitoring system and the student ID system. In both cases, the respondents have a strong preference for people over machines. They feel more comfortable with human observers and consider human observers to be a more moral choice representing less of a privacy violation. These preferences, however, vanish in the mall security system and the hotel billing system scenarios (figures 4.1c and 4.1d). Here, the preferences for humans and machines are equal. People are mostly indifferent except for the privacy violation dimension, because they consider machine observers to violate privacy more than human observers. Finally, in the airport scenario, there is a clear—albeit mild—preference for machine observers. In this scenario, people report feeling more comfortable and feel that their privacy is less violated when observed by camera systems than by security officers.

The gradient observed in these five scenarios tell us that people's tolerance of human and machine observers varies by environment. In the student ID and school attendance scenarios, people prefer human observers. This is understandable because people tend to be protective of systems that may violate the privacy of minors. Minors are vulnerable populations who may lack the ability to understand the importance—or lifelong implications—of privacy. At the same time, people are indifferent between human and machine observers in the hotel and mall scenarios, both of which are examples of large commercial settings where people expect some level of private-sector security and surveillance. Finally, we find that the preference for human observers is reversed in the airport scenario, suggesting that people in our sample may be more wary of human observers when they are backed by the coercive force of government.

Next, we explore three scenarios that move away from computer vision systems and involve data-hungry recommender systems. These include predictive purchasing, an online dating system, and a discount travel company.

 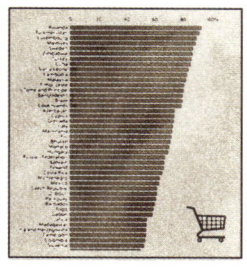 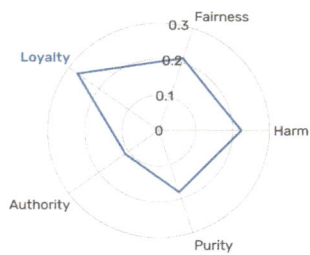

S48 A grocery delivery company announces a new service that uses data on a person's shopping habits to predict the groceries that a person will buy each week. The company assigns to each person a [dedicated shopper/AI digital twin] that uses knowledge of a person's past purchases to predict what groceries to deliver to them.

  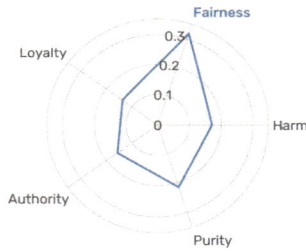

S49 An online dating system announces a new service that uses a person's past choices to set up dates for them automatically. The system requires people to make themselves available one night a week. The system guarantees a weekly blind date for them. The date is set up by a [relationship specialist/AI system].

 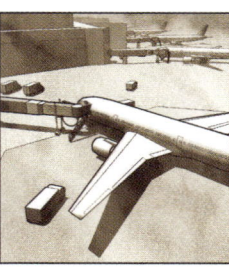 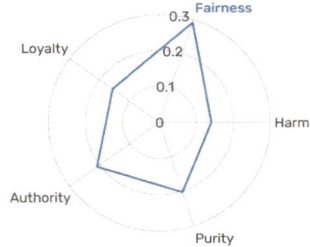

S50 An online travel company offers a discount vacation system in which users prepay a predefined amount in exchange for letting the system book a discount vacation for them. The company uses [a network of travel agents/AI system] to find and match deals with travelers' preferences.

For each scenario, participants answered the following questions (adapted to each scenario):

- Would you **recommend** the system to a friend?
(from "Would surely not recommend" to "Would surely recommend")
- Would you **trust** the decisions made by this system?
(from "Would surely not trust" to "Would surely trust")
- Would you **enroll in/use** this system?
(from "Would surely not enroll in/use" to "Would surely enroll in/use")
- Have you **ever had** groceries delivered to your home/used online dating sites/ used online traveling sites?
("No" and "Yes")
- When was the **last time** you had groceries delivered to your home/used online dating sites/used online traveling sites? (from "This week" to "More than a year ago")

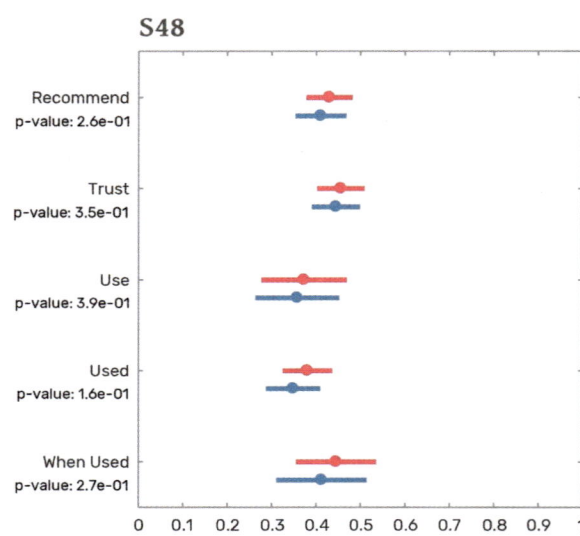

Figure 4.2

Participant reactions to three recommender system scenarios: (S48,S49,S50).

S49

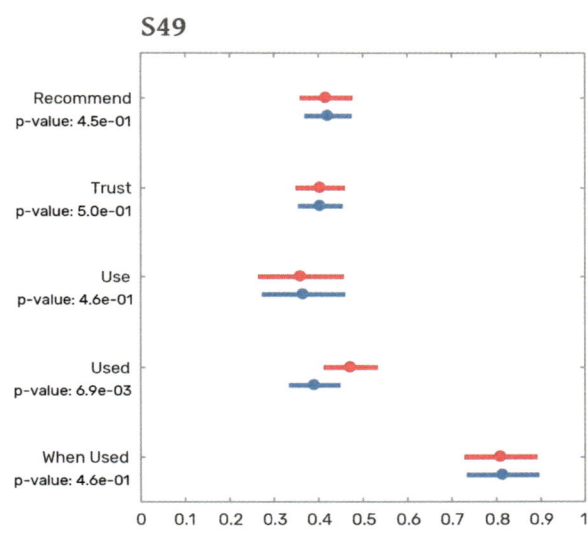

S50

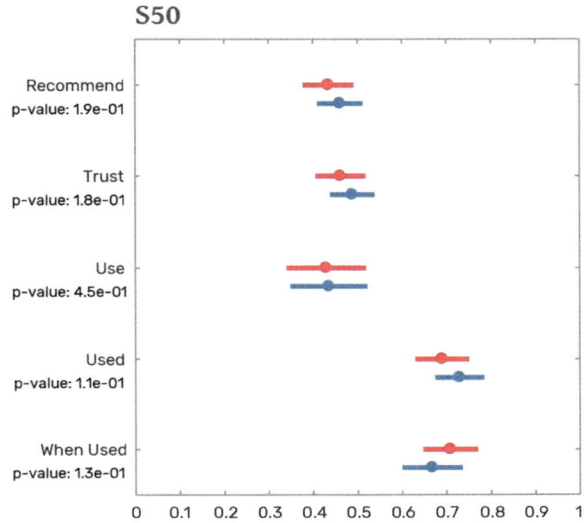

In these three examples, we find no strong preference for the use of humans or machines. This is true for systems that enjoy wide adoption today, like online dating, which more than 80 percent of the study's participants report using; or low levels of adoption, such as online groceries, which less than half report using.

Overall, our data suggest some interesting patterns. First, we find a great degree of variation among camera system scenarios. People tend to detest machine observers in scenarios involving schoolchildren and public transportation, but they are indifferent to human and machine observers in private-sector venues. This is echoed by our recommender system scenarios, which are commercial in nature and reveal no big difference between human and machine observers. The only example where we find a preference for machine observers is the airport scenario, which suggests an interaction between human observers and the coercive power associated with governments. To explore that relationship further, consider the following citizen scoring scenario, which was evaluated using the same questions used for the computer vision scenarios.

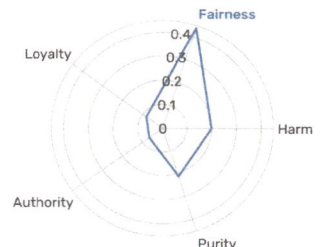 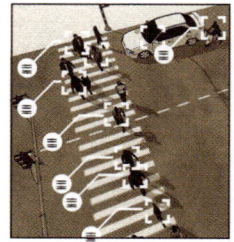 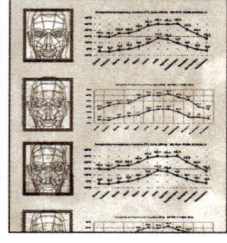

S51 To improve citizen behavior, a party proposes to implement a scoring system for each citizen. The system is based on [a hotline where citizens can anonymously report others/AI and big data]. The scoring system is used to determine people's creditworthiness, grant admission to public universities, and for hiring and promotion in government jobs.

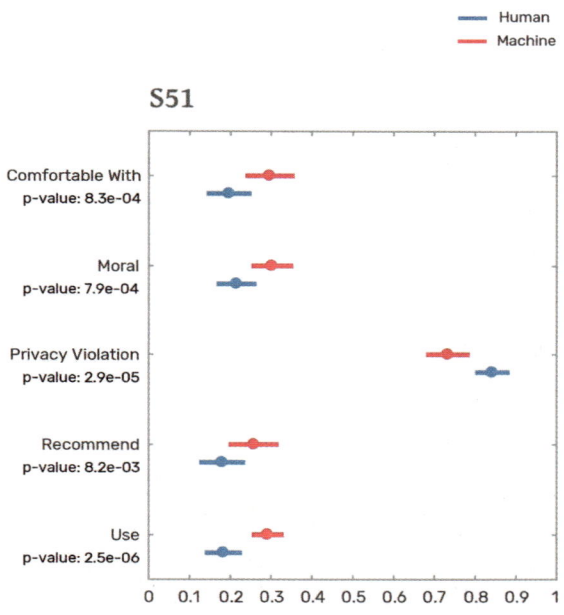

Figure 4.3

Participant reactions to the citizen scoring scenario.

Figure 4.3 shows the results for the citizen scoring scenario. Overall, people reject the idea of citizen scoring, but they do so more strongly when this is implemented in systems that involve people telling on each other than on systems based on algorithms and big data. This aligns with our observation for the airport scenario, but it affords multiple interpretations. On the one hand, having a system where people are incentivized to tell on others has perverse incentives: people could report others not because they've done anything wrong, but because they are rivals or enemies.

Machines are not expected to have such vindictive motives, and hence are less likely to have this perverse incentive. On the other hand, people could be reacting to people telling on others because there are social norms against *ratting out*. Reputation is important to people, and social norms tell us to think twice before we try to ruin another's reputation. Moreover, this scenario—like that of the airport—also involves the coercive power of the state, so this could be yet another interpretation of why people dislike the mechanical approach less.

In this chapter, we compared people's reactions to machine and human observers in a variety of settings. We found that people's preference for human and machine observers varies across scenarios. Yet our results are agnostic about the mental models that people have of machine and human observers. Would people's preference for machine and human observers change if we explicitly described the privacy-preserving protocols involved? Studies about people's attitudes toward randomized response suggest so.[22] But for the time being, this is the end of our journey. In the next chapter, we move away from privacy to focus on another fear induced by machines: the fear of labor displacement.

Working Machines

5

CHAPTER 5

In 2014, the *New York Times* reported a story about Jannette Navarro, a mother of a 4 year old who at the time was working at a Starbucks in Southern California.[1] Navarro not only had to battle a 3 hour commute, but she also "rarely learned her schedule more than three days before the start of a workweek." The unpredictability of her schedule bordered on cruelty. She was asked to "work until 11 p.m. on Friday [and] report again at 4 a.m.," a practice that workers like her knew as *clopening*. Navarro's unpredictable work schedule made her life incredibly complicated. Finding someone to take care of a 4 year old is challenging, but it is especially hard when you are constantly required to do so with only a few days' notice. But Navarro's schedule was not being prepared by a sadistic manager. It was made by an algorithm created by a company called Kronos, a vendor that Starbucks hired to optimize its labor force.

Starbucks updated its practices immediately after the *Times* ran Navarro's story.[2] Yet practices such as clopening still prevail in the low-wage sectors of the US economy.[3] For the purpose of this chapter, however, Navarro's story illustrates two important aspects of the effects of technology on labor. The first is the simple idea of labor displacement, which is embodied in the fact that Navarro's schedule was not being managed by a human, but by an algorithm. The second is the idea that technology can decrease the quality of work, an effect known as *precarization*, which is defined as reducing the material and psychological welfare of a job.

In recent years, people have grown concerned about technological labor displacement and the precarization of work.[4] But these concerns are not new.[5] Concerns about the influence of technology on labor are as old as the introduction of printing in Europe. As printing spread,[6] monastic scribes attempted to ban presses, declaring them demonic devices.[7] Centuries later, English Luddites became famous for opposing steam-powered looms. But the rage of Luddites was not only about labor displacement. It was the abysmal labor conditions of the Industrial Revolution, a clear example of the precarization of work.[8]

In the twentieth century, fears of automation took over the public dialogue in the US at least twice. In the 1960s, fears of technological displacement grew after *Time* magazine published a popular article in 1961 on "The Automation Jobless."[9] "Not Fired, Just Not Hired," the subhead continued, building a case on the effects of technology on the future of labor.

Recently, displacement fears revived, together with reports claiming that almost half of all jobs could be automated [10] and that this change could be happening "ten times faster [than] and at 300 times the scale" of the Industrial Revolution.[11] Yet most of the academic literature on labor and automation has embraced a less alarmist approach.[12]

Labor economists have been eager to emphasize that technology is not only a substitute for labor, but also a complement,[13] so it creates jobs with one hand and takes them away with the other.[14] Economists agree—in general—that technology is labor saving,[15] but many also say that it increases the productivity of the workers that it does not replace. These increases in productivity, plus new complementarities, can increase aggregate demand and stimulate the need for more human work.

A classic example of the complex interaction between technology and labor is the introduction of automatic teller machines (ATMs). ATMs did not eliminate human tellers, as some feared. In fact, the number of human tellers in the US actually grew modestly after ATMs were introduced, from 500,000 in 1980 to about 550,000 in 2010.

ATMs did not eliminate the job of teller; they transformed it. This was in part due to the lower cost of opening new bank branches, which together with other factors, such as more bank-friendly regulations, contributed to new bank teller jobs with different responsibilities.[16]

A more modern example of the complex interaction between technology and labor can be found in China. In cities like Nanjing, it is common for restaurants to have QR codes on every table. The QR code allows customers to order food and pay their bills using their phones. But this technology does not replace the need for human servers. It only automates a few of their tasks, allowing them to focus on things other than taking orders or collecting checks. Servers are still needed to carry food, clear tables, greet customers, deal with special requests, and maintain a civilized environment at the restaurant. This example also shows that automation often does not replace entire jobs because it involves tasks. That is why studies that focus on the automation of jobs tend to overestimate the impact of automation[17] compared to studies focused on the automation of tasks.[18]

Hence, the question that we should be asking is not "Will a robot will take my job?" but "How will the labor market change with technology?" In response to that question, economists have made a few predictions.

On the one hand, there is an apparent consensus that while changes in technology have important effects on labor on the short term, they do not appear to affect the need for labor in the long run.[19] Using data from the International Federation of Robotics (IFR), Graetz and Michaels report that between 1993 and 2007, the introduction of robots did not reduce total employment, although they do find evidence that robots reduce the employment share of low-skilled workers.[20] Other authors also find a negative correlation between the stock of robots in a country and unemployment.[21]

On the other hand, there is no clear consensus on predictions about the redistributive effects of technology. Some scholars anticipate an increased polarization of the labor force and increased inequality.[22] Yet some scholars have arrived at the opposite conclusion when focusing on the replacement of tasks rather than occupations.[23]

Another angle of this discussion has been to focus on the types of jobs being replaced by new technologies. In his book *Prediction Machines*, Ajay Agrawal focuses on the fact that artificial intelligence (AI) technologies are mostly good for prediction,[24] so he forecasts the effect of AI on labor by assuming that its main effect is a reduction in the cost of predictions. For instance, lower prediction costs could flip the shopping-then-shipping model of online retailers to a shipping-then-shopping model. This is because, in a world where stores can predict the items that a person may buy, business models in which stores ship items and learn from the ones that are returned may become viable.

The fact that technology will affect the future of work is undeniable. But technology is not the only force affecting labor. The future of work also depends on global value chains,[25] the increasing concentration of complex economic activities,[26] the rising education levels of the Global South,[27] and international migration.[28] To better manage this impact, we need to understand how people react to the impact of technology on jobs compared to other forces.

The goal of this chapter is to compare people's reactions to displacement attributed to technology with displacement attributed to humans. We contribute to that goal by using two sets of scenarios. The first set compares technology-based displacement with displacement attributed to foreign temporary workers. The second set compares technology-based displacement with displacement coming from four sources: foreign temporary workers, foreign contractors (outsourcing), foreign subsidiaries (offshoring), and younger workers. Let's begin by looking at the first set of examples.

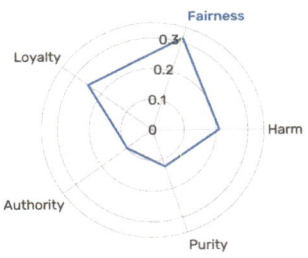   

S52 A trucking company is looking to lower costs by bringing in [temporary foreign drivers/autonomous trucks]. This change reduces the company's costs by 30 percent, but several local drivers lose their jobs.

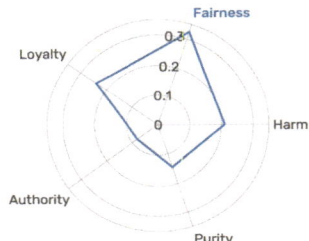 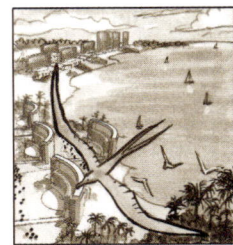  

S53 A large chain of luxury resorts decides to lower the cost of staffing their poolside bars by bringing in [temporary foreign workers/vending and cooking robots]. The [workers/robots] can take a guest's room number for payment purposes and serve a large variety of cocktails and dishes. As a result of the change, several local workers lose their jobs.

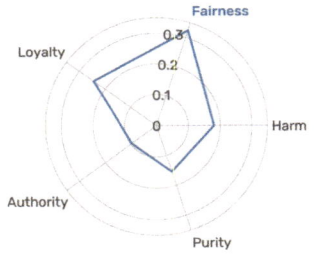   

S54 A nuclear power plant is looking to lower their operational costs. They decide to [bring in foreign nuclear technicians/buy an AI operation system]. This change allows the company to reduce their operational costs by 30 percent, but several local technicians lose their jobs.

   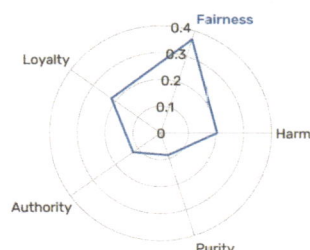

 A school is looking to lower their costs by [bringing in foreign teachers/adding robot teachers to some of their classes]. As a result, the school reduces its costs by 30 percent but fires several local teachers.

For each scenario, we asked the following questions:

- Do you **approve** of this change? (from "Strongly disapprove" to "Strongly approve")

- Would you **ban** this change? (from "Would surely not ban" to "Would surely ban")

- How **morally** wrong or right was the action of the manager? (from "Extremely wrong" to "Extremely right")

- Does your **opinion** of this organization worsen or improve because of this change? (from "Worsens extremely" to "Improves extremely")

- Do you think **others will approve** of this change?
  (from "Strongly disapprove" to "Strongly approve")

- If you were in a similar situation as the manager, would you have done the same? (from "Definitely not" to "Definitely yes")

Figure 5.1 compares technological displacement with displacement attributed to foreign workers in these four scenarios. In general, we find a preference for displacement attributed to technology over displacement attributed to foreign workers. The strength of this preference, however, varies depending on the scenario. In the case of trucking, the preference to ban foreign workers is much stronger than the preference to ban autonomous trucks. People also approve of foreign workers less than autonomous vehicles and think that others will have similar opinions. The resort scenario shows a similar pattern, albeit with lesser differences.

These differences persist, although weakened, in the power plant scenario, and they vanish altogether in the school scenario. We should note, however, that in the school scenario, people have a strong preference against both machines and foreigners replacing teachers. Thus, the fact that we do not observe strong differences may be due to a floor effect among two unpopular options.

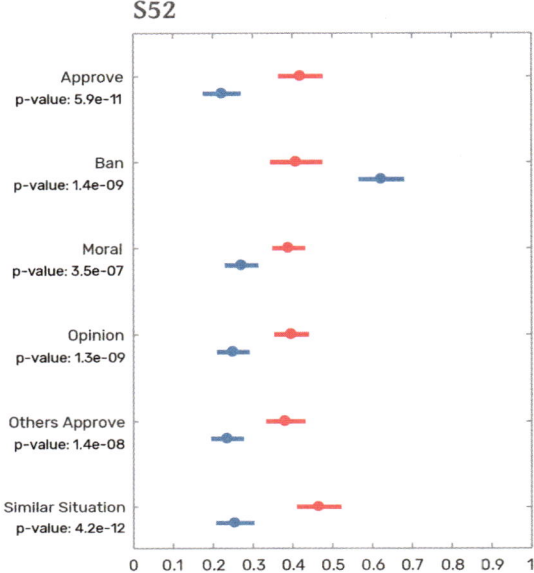

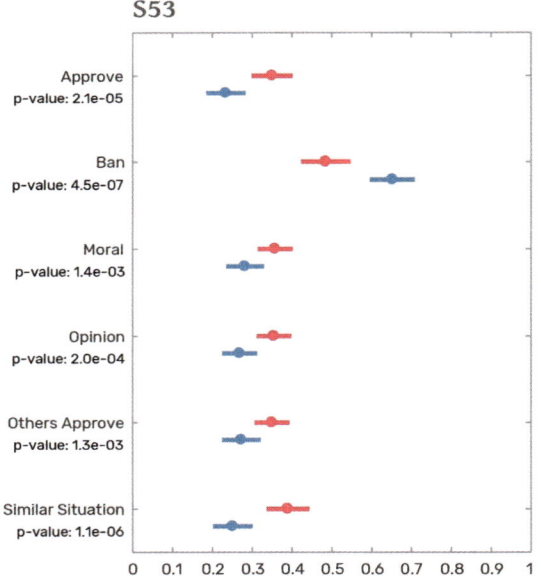

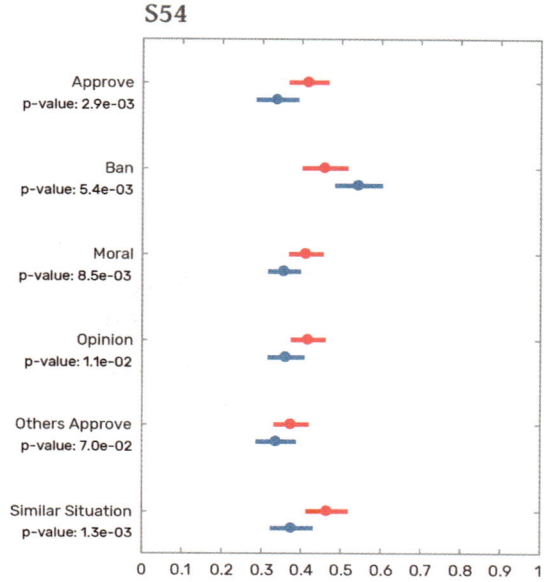

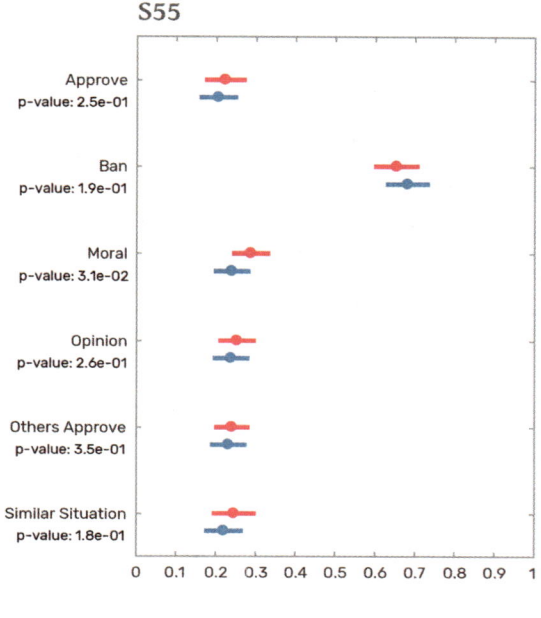

Next, we look at scenarios comparing labor displacement with displacement due to other sources, including foreign temporary workers, foreign contractors (outsourcing), foreign subsidiaries (off-shoring), and younger workers.

Once again, we find results that are quite consistent across all scenarios (see figure 5.2). People tend to approve more and are less willing to ban displacement caused by technology than displacement involving other people.

People react particularly strongly against displacement attributed to foreign workers and replacing older workers with younger workers. Among the nontechnological forms of displacement, people react less negatively to opening a foreign subsidiary (offshoring), followed by hiring a foreign contractor (outsourcing).

Figure 5.1

Participant reactions to displacement by foreign workers, as opposed to technology, in four scenarios: (S52,S53,S54,S55)

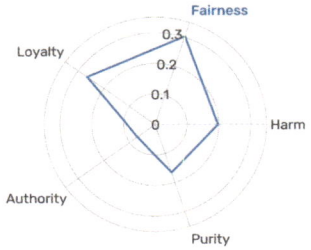

S56 A law firm is looking to lower their costs for routine clerical work. They decide to [open a branch in a low-income country/hire a foreign contractor/bring in foreign workers with temporary visas/replace older workers with younger workers/buy an AI legal system]. The result is a reduction in costs and the firing of several of their local staff.

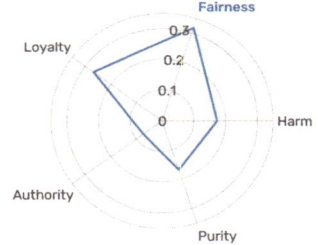

S57 A software firm is looking to lower the costs of their routine maintenance and updating tasks. They decide to [open a branch of the firm in a low-income country/hire a foreign contractor/bring in foreign workers with temporary visas/replace older workers with younger workers/buy an AI system]. The result is a reduction in costs and the firing of several of their local staff.

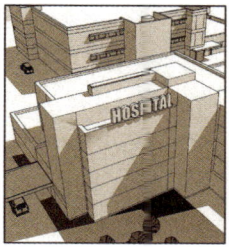   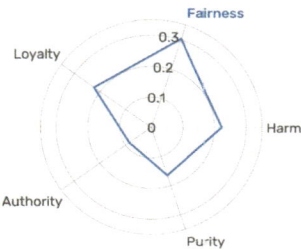

S58

A hospital is looking to lower their diagnostic costs for X-rays and computerized axial tomography (CAT) scans. They decide to [open a branch in a low-income country/hire a foreign contractor/bring in foreign workers with temporary visas/replace older workers with younger workers/buy a computer vision system]. The result is a reduction in costs and the firing of several of their local staff.

   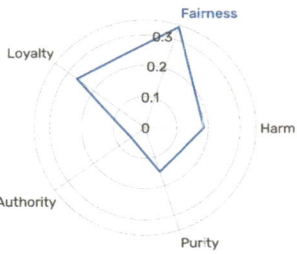

S59

A manufacturing company is looking to lower their production costs. They decide to [open a plant in a low-income country/hire a foreign contractor/bring in foreign workers with temporary visas/replace older workers with younger workers/buy a robotic manufacturing system]. The result is a reduction in costs and the firing of several of their local staff.

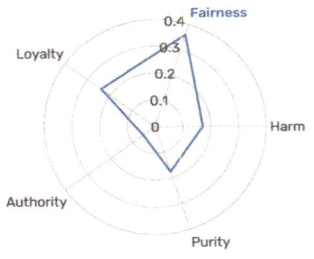   

S60 A film studio is looking to lower their animation costs. They decide to [open a studio in a low-income country/hire a foreign contractor/bring in foreign workers with temporary visas/replace older workers with younger workers/buy AI animation software]. The result is a reduction in costs and the firing of several of their local staff.

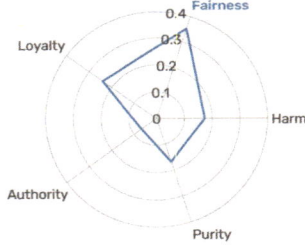   

S61 A finance company is looking to lower their fund management costs. They decide to [open a branch in a low-income country/hire a foreign contractor/bring in foreign workers with temporary visas/replace older workers with younger workers/buy AI investment software]. The result is a reduction in costs and the firing of several of their local staff.

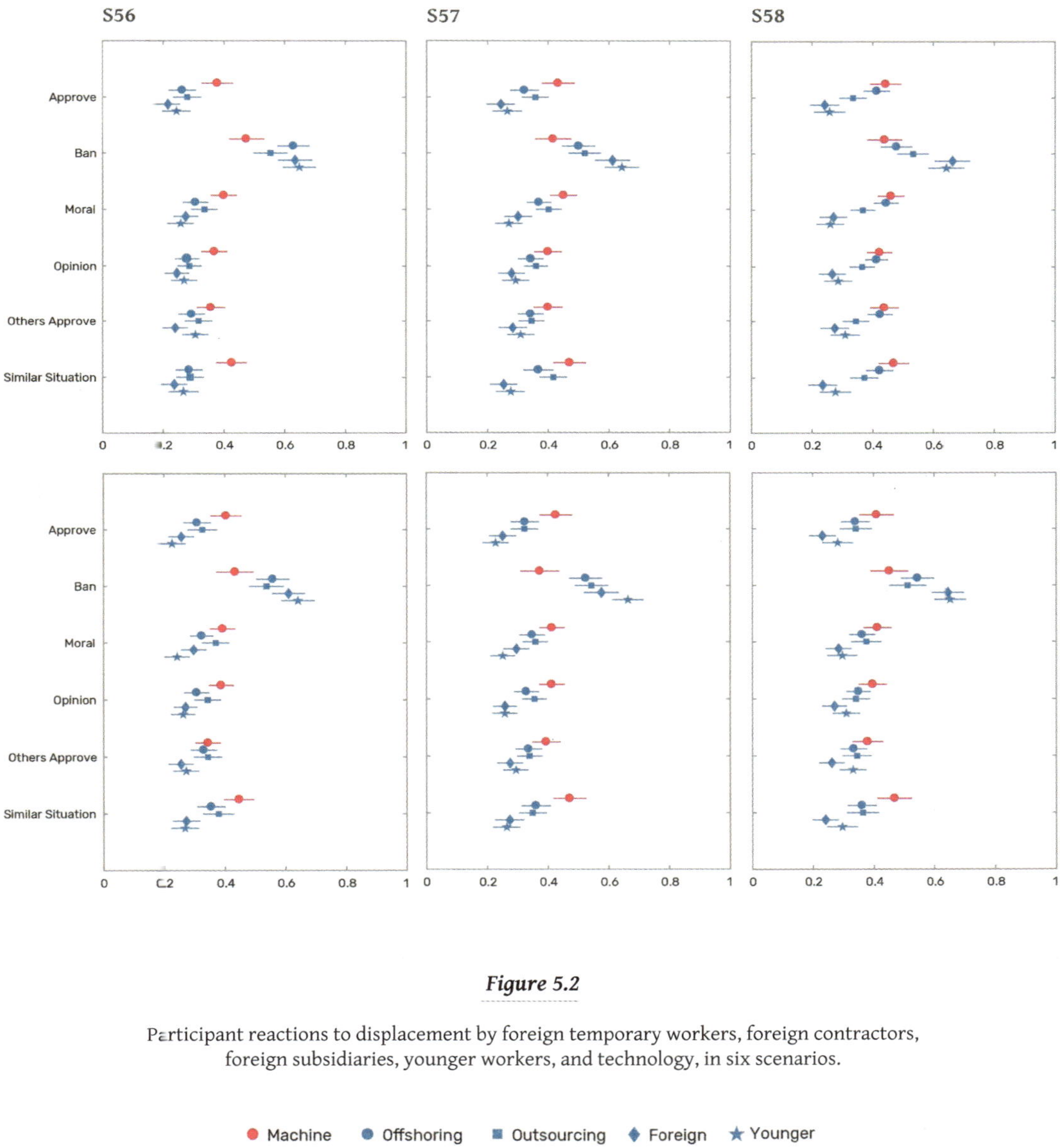

S56

S57

S58

Figure 5.2

Participant reactions to displacement by foreign temporary workers, foreign contractors, foreign subsidiaries, younger workers, and technology, in six scenarios.

● Machine ● Offshoring ■ Outsourcing ◆ Foreign ★ Younger

Together, these scenarios show that people tend to react less negatively to technology-based displacement than to displacement based on other humans. This may be the case for a variety of reasons.

First, people may see technological displacement as more inevitable. People may see competing against a machine designed to excel at a specific task as futile, but competing against other humans, even when they are younger or foreign, as always possible. A second possibility is that the negative reactions against displacement by foreign workers are automatic responses to well-socialized "in-group versus out-group" biases. In the US, displacement by foreigners is a narrative with a well-established negative connotation. Also, people may perceive displacement by foreigners and younger people as more imminent to them, especially if they or someone they know has experienced a similar situation.

Third, it could be that people oppose cost reductions based on cheaper labor more strongly because they consider profiting from lower salaries to be more exploitative, and less acceptable, than profiting from technology. In fact, when we look at the moral dimensions associated with these scenarios, we find that they trigger the fairness dimension of moral psychology. As we saw in chapter 3, people tend to react more strongly to humans in situations that they perceive as unfair, so this could be yet another effect that contributes to explaining our observations. Finally, people could see replacement by cheaper labor as retrograde compared to replacement by technology, which could be seen as progress.

Regardless of the explanation, what is true is that within this sample, there are clear negative reactions to labor displacement, which are amplified when displacement is attributed to foreign or younger workers, but which are still there for displacements attributed to machines. Because of these reactions, it is not surprising to find work focused on mitigating the potential negative consequences of technological displacement. Some of these alternatives have a strong taste for regulation, while others focus more on additional market flexibility.

On the side with a stronger taste for regulation, we find people in favor of a robot tax (i.e., a tax on the profits of companies that use more robots). The argument is that because most tax revenue comes from labor income, tax policies tacitly incentivize automation.[29] By adopting automation, companies reduce their labor costs, as well as the taxes they pay on their employees. In this view, automation erodes the overall tax base if robotic labor goes untaxed. Of course, there are some clear counterarguments to this line or reasoning. For instance, if automation does not cause unemployment but simply shifts workers to different jobs, we cannot use this argument to justify a robot tax. Also, robot workers do not consume government services in the same way that human workers do, so their tax bill would not need to cover for items like pensions, health care, and education, which taxes on labor usually cover.

On the side arguing for more flexibility, we find proposals focused on removing barriers limiting the ability of workers to move between occupations, and limiting new business models from entering established sectors. One barrier to labor mobility is the excessive need for state licensing. In the US, the need for a state license has grown from 5 percent to 30 percent of workers between 1950 and 2008. As McAfee and Brynjolfsson argue, "Some of the requirements are plainly absurd: in Tennessee, a hair shampooer must complete 70 days of training and two exams, whereas the average emergency medical technician needs just 33 days of training."[30] Labor mobility is known to be an important channel of knowledge diffusion,[31] so barriers for workers to move between occupations and industries, like excessive licensing, can reduce the ability of the market to adapt to changes in technology.

On that front, Alan Krueger and Seth Harris[32] advocate for a new worker classification that sits somewhere between full-time employees and contractors: "independent workers." Independent workers "would not be eligible for overtime pay or unemployment insurance, but would enjoy the protection of federal antidiscrimination statutes and [would] have the right to organize, . . . withhold taxes and make payroll tax contributions."

Finally, between both of these camps, we have the idea of universal basic income (UBI). UBI is not a new idea. It can be traced back to Thomas More and Condorcet.[33] Still, UBI is an idea that has recently regained popularity. One of the modern versions of UBI is the idea of paying a guaranteed income to all citizens, which they can then use to procure services that are often the purview of government, like education, health care, or affordable housing. On the one hand, UBI is quite market oriented, as it entrusts basic social safety nets to cash transfers and the market. On the other hand, the source of these funds is public, making it more of a government intervention. Not surprisingly, UBI is a divisive topic, with some arguing that it is excessive and impractical,[34] and others— like recent presidential candidate Andrew Yang—touting it with enthusiasm.[35]

In this chapter, we compared people's reactions to scenarios involving labor displacement attributed to either technology or humans. We found that, on average and across most scenarios, people reacted less negatively to technological displacement. They were less prone to ban it and accepted it more than other forms of displacement, especially when the displacement was attributed to foreign or younger workers.

In chapter 6, we will zoom out from individual scenarios and look across them instead. We will use data from all the cases studied so far, as well as data from additional ones (shown in the appendix), to discover trends and patterns that are hard to observe by looking at scenarios alone. By zooming out, we will lose granularity but gain the power of abstraction in our quest to understand how humans judge machines.

Moral
Functions

6

CHAPTER 6

Imagine designing a machine to mimic the moral judgment of humans. In principle, you may want a machine that is better than humans at making moral judgments. But in practice, that goal may be too farfetched. So, instead, you may want to first make a machine that simply mimics the moral judgment of humans.

The goal of this chapter is to explore the very basics of that machine. To achieve that goal, we will use simple statistical tools that prioritize explicability over accuracy. These tools will help us zoom out of individual scenarios by providing descriptions that are less nuanced, but more generalizable. They will also inform us about the impact of different inputs into moral judgments.

Our exploration will build on the idea of a *moral function*: a mathematical object predicting how people will judge the outcomes of a moral scenario based on inputs, such as who is performing the action, or its level of perceived harm. One input that is of particular interest for us is whether the agent performing the action in a scenario is a human or a machine. Throughout the book, we have seen that people judge human and machine actions differently. This is consistent with the social psychology literature telling us that people judge and punish more severely members of out-groups (in our case machines) than members of the in-group (in our case, humans).[1,*]

By using moral functions, we can formalize those differences by exploring how they relate to the characteristics of a scenario.

Our approach will rely on many simplifying assumptions,[†] which we introduce in an effort to prioritize clarity. To make that explicit, we will mention the problems caused by these simplifying assumptions when we introduce them.

To begin, we introduce the *moral space* a quantitative representation of moral judgment. This representation, which we use to abstract away from the details of each scenario, is inspired by Jonathan Haidt's moral foundation theory[2] and is based on three factors: harm, intention, and wrongness. While in principle, we could include many inputs, such as whether the dilemma involves an uncertain outcome or represents

---

[*] This intergroup bias develops as children grow, and as such, it can be detected as soon as six years old (J. J. Jordan, K. McAuliffe, and F. Warneken, "Development of In-group Favoritism in Children's Third-Party Punishment of Selfishness," PNAS 111 (2014): 12710–12715). Moreover, neuroimaging research shows that people have higher sensitivity (i.e., great activity in the left orbitofrontal cortex) to the suffering of in-group members than out-groups when an out-group member performs the harmful action. (P. Molenberghs, J. Gapp, B. Wang, W. R. Louis, and J. Decety, "Increased Moral Sensitivity for Outgroup Perpetrators Harming Ingroup Members," Cerebral Cortex 26 (2016): 225–233). In an experiment in which Swiss army officers played a prisoner's dilemma, researchers found more cooperation among officers from the same platoon and harsher punishments for defectors from different platoons. (L. Goette, D. Huffman, S. Meier, and M. Sutter, "Group Membership, Competition, and Altruistic Versus Antisocial Punishment: Evidence from Randomly Assigned Army Groups," IZA Discussion Paper No. 5189 (2010), https://papers.ssrn.com/abstract=1682710.) When asked to imagine a theft, undergraduate students assigned higher fines to foreign offenders than to relatives or classmates (D. Lieberman and L. Linke, "The Effect of Social Category on Third Party Punishment," Evolutionary Psychiatry (1 April 2007). Similar patterns have been observed for affiliations with soccer clubs and political parties, (B. Schiller, T. Baumgartner, and D. Knoch, "Intergroup Bias in Third-Party Punishment Stems from Both Ingroup Favoritism and Outgroup Discrimination," Evolution and Human Behavior 35 (2014): 169–175.) and even among tribes in Papua New Guinea. (H. Bernhard, U. Fischbacher, and E. Fehr, "Parochial Altruism in Humans," Nature 442 (2006): 912–915).

[†] Our presumption is that all the statistical estimates presented here can be improved, but more sophisticated estimation techniques may obscure or distract from the key concepts that we want to communicate.

a violation of a moral dimension other than harm, we focus for simplicity only on five variables: the perceived levels of harm, intention, and wrongness of a scenario, and whether the scenario was a treatment or a control (i.e., whether the action was performed by a human or a machine). We then explore how the characteristics of the respondents—the people judging the scenarios—affect moral judgments.

In this representation, each scenario is described by two dots connected by a line. The red dot shows the judgment of the machine action, while the blue dot shows the judgment of same action when conducted by a human. The dots exist in a three-dimensional space defined by wrongness on the vertical axis (the z-axis) and harm and intention on the horizontal plane (the x- and y-axes).

Figure 6.1 illustrates the simplified moral space using average answers for perceived wrongness, harm, and intention. The black line connecting the dots shows that both dots belong to the same scenario. We use a diverging scale for wrongness, meaning that wrongness values range from "Extremely right" (0) to "Extremely wrong" (1), with the neutral value ("Neither wrong nor right") at 0.5. For harm and intention, we use a sequential scale. That is, intention ranges from "Not intentional at all" (0) to

Figure 6.1

Quantitative representation of judgments observed for human and machine actions in a scenario.

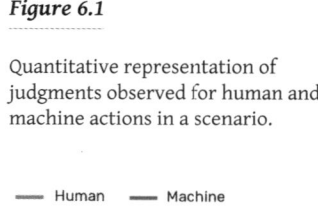

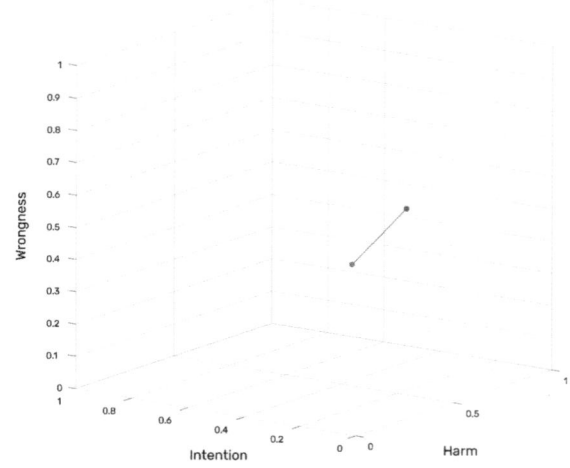

"Extremely intentional" (1). Similarly, harm ranges sequentially from "Not harmful at all" (0) to "Extremely harmful" (1).

We can use this representation to summarize the patterns found across all the scenarios that included questions on perceived wrongness, harm, and intention. This excludes the privacy and labor displacement scenarios, which did not include these three questions.

Figure 6.2 shows a summary of our experimental results. Note that the moral space is purely descriptive, which allows us to consider wrongness, harm, and intention simultaneously, even though these are all affected by the treatment.

The first finding, which is interesting but slightly obvious, is that moral judgments do not populate the whole space. They fall within a plane that extends from the upper-left corner, with high levels of harm, wrongness, and intention, to the right side of the cube, which shows scenarios with low levels of wrongness and harm. This is because some corners, such as scenarios with no intention or harm, cannot be high in wrongness. Similarly, scenarios high in harm and intention cannot be rated as low in wrongness.

These constraints limit the observation to relatively narrow moral planes. In the next section, we will model these planes mathematically. In this section, we explore the patterns found in this three-dimensional space by looking at the three faces of the cube separately.

Figure 6.3 zooms into the harm-intention plane. Here, we see that machine actions are seen as less intentional than human actions when the level of human intention is relatively high, which is true for most cases in our sample. However, we find six scenarios in which the actions of machines are seen as more intentional than those of people. These six cases are all at relatively low levels of intention and include the excavator scenario (S1), the wrong national anthem scenario (S17), the school demolition scenario (S18), and the four car accidents scenario (S11–S14).

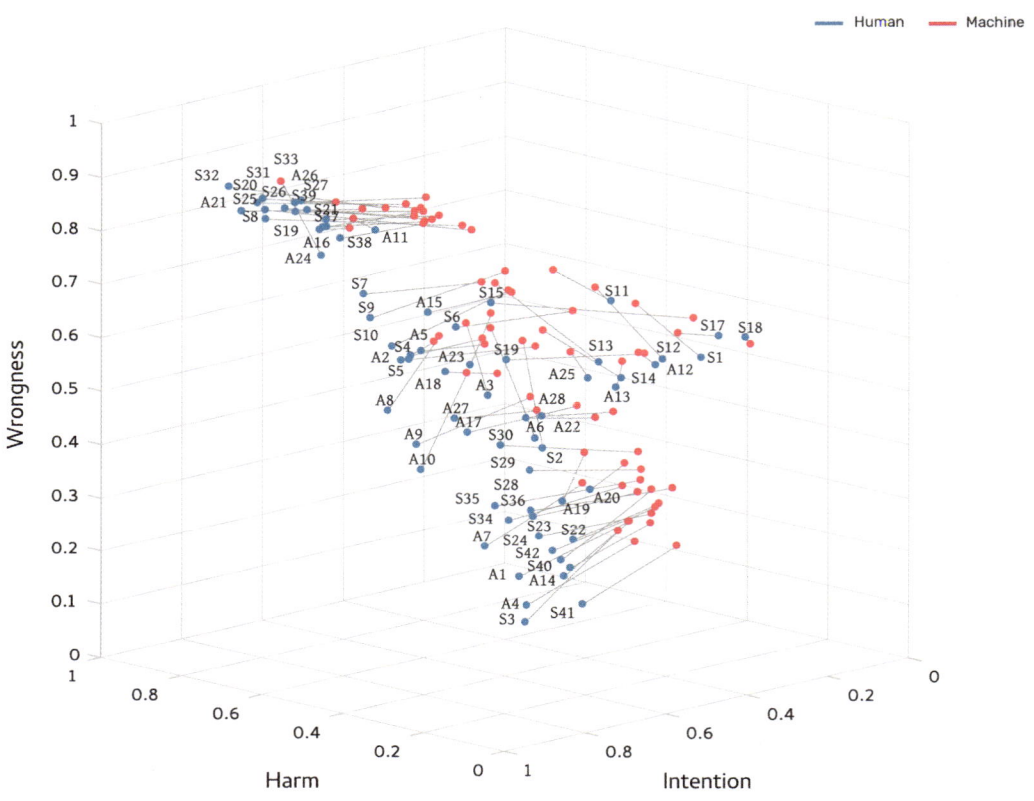

Scenarios

S1: Graveyard
S2: Tsunami effort fails
S3: Tsunami effort succeeds
S4: Tsunami effort half succeeds
S5: Ad with kiss
S6: Ad with play on "riding"
S7: Ad with clothes addiction
S8: Record label plagiarism
S9: God comedy sketch
S10: Performance art piece
S11: Sunny day accident with person
S12: Sunny day accident with dog
S13: Windy day accident with person
S14: Windy day accident with dog
S15: Cleaner with national flag
S16: Anthem interruption
S17: Wrong anthem
S18: Demolished school flag
S19: Hispanic HR discrimination

S20: African American HR discrimination
S21: Asian HR discrimination
S22: Hispanic HR fairness
S23: African American HR fairness
S24: Asian HR fairness
S25: Hispanic college discrimination
S26: African American college
   discrimination
S27: Asian college discrimination
S28: Hispanic college fairness
S29: African American college fairness
S30: Asian college fairness
S31: Hispanic salary discrimination
S32: African American salary
   discrimination
S33: Asian salary discrimination
S34: Hispanic salary fairness
S35: African American salary fairness
S36: Asian salary fairness

S37: Hispanic police discrimination
S38: African American police
   discrimination.
S39: Asian police discrimination
S40: Hispanic police fairness
S41: African American police fairness
S42: Asian police fairness
A1: Fire effort success
A2: Fire effort half success
A3: Fire effort fail
A4: Good prediction
A5: Hurricane effort half success
A6: Hurricane effort fail
A7: Supermarket robbery
A8: Pharmacy robbery
A9: Jewelry robbery
A10: Bank robbery
A11: Ambush
A12: Raw material shortage

A13: Excess material
A14: Good prediction
A15: Counseling
A16: Hiking
A17: Company brand names
A18: Amusement parks
A19: Ray's painkillers
A20: Ben the physical therapist
A21: Twitter
A22: Gas tax
A23: Interest rates
A24: Suspected terrorist
A25: Senior assistant
A26: Civil engineer collapse
A27: Civil engineer success
A28: Shoplifting security

Scenarios beginning with an A can be found in the appendix

Figure 6.2

Judgments of human and machine actions across scenarios.

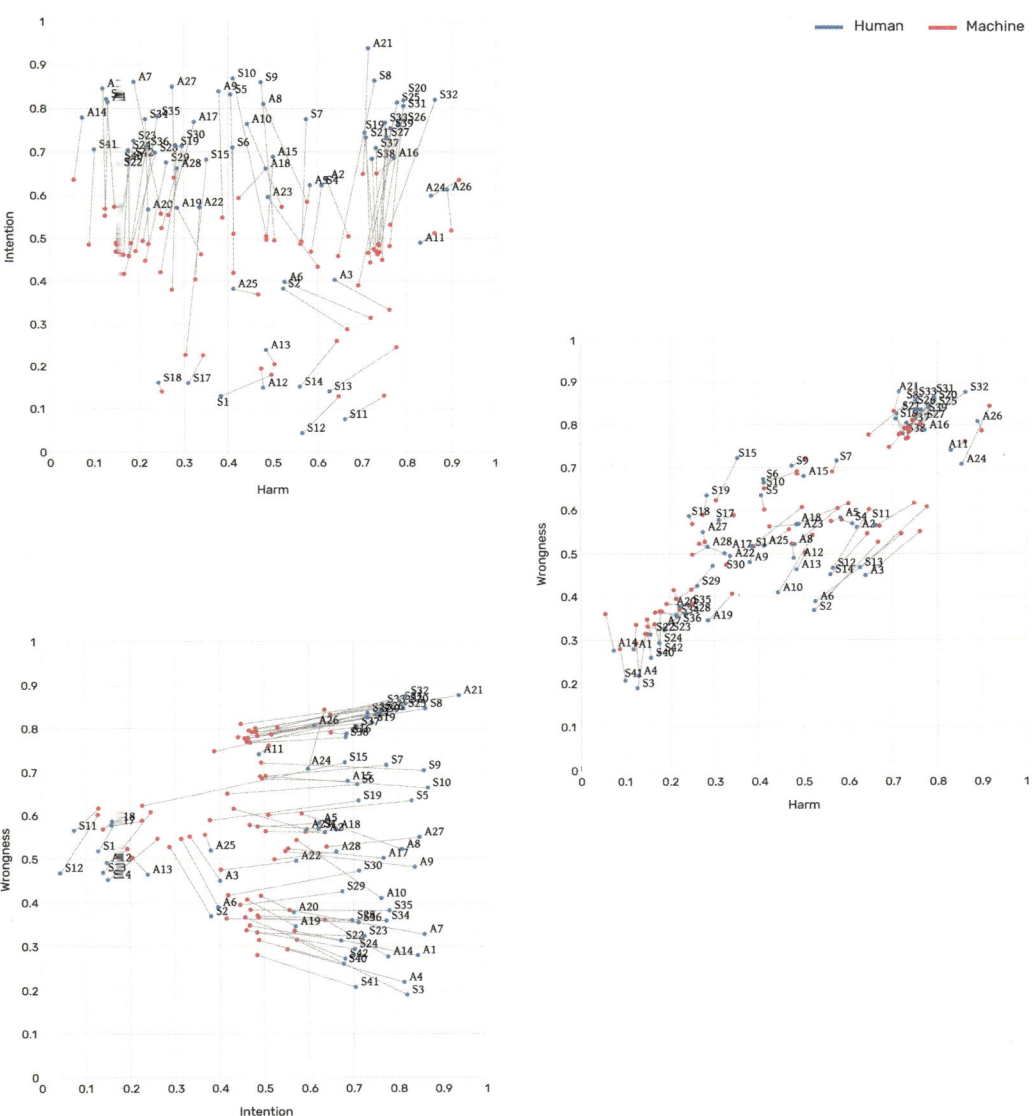

Figure 6.3

Harm-intention plane, wrongness-intention plane, and harm-wrongness plane.

The harm-intention plane reveals two things: The first, which is obvious, is that in most cases, people appear to assign more intention to human actions than machine actions. The second, which is more surprising, is that people may excuse human actions more than machine actions in accidental scenarios. For instance, when a car accident is caused by either a falling tree or a person jumping in front of a car, people assign more intention to the machine than to the human behind the wheel. As discussed in previous chapters, this suggests that people perceive an accident more like an error when the actor is a machine, but as misfortune when the actor is a human. Hence, in these types of scenarios, they forgive or excuse humans more than machines.

Figure 6.3 also shows the wrongness-intention plane. We also see a triangular pattern because intention modulates the level of perceived wrongness. Unintentional actions cluster close to the neutral value (0.5) "Neither wrong nor right." But actions perceived as intentional can score very high ("Extremely wrong") or very low ("Extremely right"). This is consistent with an extensive body of literature in moral psychology showing that intentional actions are judged worse than accidents, even when the accidents have more serious consequences.[3]

But the wrongness-intention plane also reveals some interesting patterns. For low levels of intention ($I < 0.3$), we see a clear upward slope, meaning that machine actions are perceived as both more wrong and more intentional than those of humans. This group contains the four car accident scenarios (S11–S14).

At an intermediate level of intention ($0.3 < I < 0.4$), we find actions that are perceived as less intentional for machine, but also worse. These examples include those of unlucky decisions under uncertainty, like the tsunami scenario (S2), or cases with equivalent outcomes for the fire and hurricane framings (A1 and A4).

At high levels of intention, however, differences in the intention attributed to humans and machines correlate with differences in the level of perceived wrongness. For high wrongness (> 0.75), human actions are judged as more intentional and more morally extreme (worse). This group consists of cases involving discriminatory treatment in school admissions and human resources (S19–S21, S25–S27, S31–S33, S37–S39). For low wrongness (< 0.4), machine actions are seen as less intentional, but still are judged worse than the equivalent action performed by a human. This group includes cases such as those involved in correcting unfair treatment in school admissions and human resources (S19–S21, S22–S24, S25–S27, S34–S36, S40–S42). In other words, because human actions are seen as more intentional, humans are perceived as more morally right than machines in scenarios with strong positive outcomes, and as more morally wrong than machines in scenarios with strong negative outcomes.

We look at the harm-wrongness plane (figure 6.3). Unsurprisingly, we see a strong positive correlation between perceived harm and perceived wrongness. Yet we also observe regions characterized by different regimes. For positive outcomes (W < 0.35), we find no big difference between the harm attributed to a machine or a human action, but we do find that machine actions are judged worse. At intermediate levels of harm and wrongness (0.4 < H < 0.75 and W < 0.65), we find actions that are perceived as more harmful and worse when performed by machines than humans. In fact, the evaluation of these scenarios is so extreme that humans are—on average—perceived to be morally right (W < 0.5) in situations in which machines are perceived—on average—as morally wrong (W > 0.5). In this region, machines are also perceived as more harmful. This cluster is populated by accidental scenarios, including the car scenarios (S11–S14), the interest rate scenario (A23 in the appendix), and the unlucky outcome of the tsunami scenario (S2). In these uncertain cases, people are less forgiving of machines and judge actions as more harmful and morally worse when they are performed by machines.

Finally, for scenarios rated high on harm and wrongness (W and H > 0.7), we find two groups. The first one involves cases of algorithmic bias (chapter 3), which relates to the fairness dimension of moral psychology. Here, human actions are seen as both slightly more harmful and also worse than the equivalent actions performed by a machine. The second group, which exhibits the opposite trend, consists of two cases of accidental manslaughter, such as the terrorist scenario (A24) and the ambush scenario

(A11). Here, machine actions are seen as more morally wrong than those of humans, suggesting once again that the bias against machines is modulated by a scenario's moral dimensions.

The moral space tells us that the way in which people judge the actions of machines compared to those of humans varies across scenarios. When intention and harm are low, people appear to be less forgiving of machines, evaluating their actions as worse. When intention and harm are high, however, people tend to judge human actions as worse than the equivalent machine actions.

Of course, the results presented here should be taken with a grain of salt. Despite the apparent clarity of these trends, the moral space should include factors beyond a scenario's perceived level of harm and intention. For instance, in scenarios involving a dimension of fairness, such as the algorithmic bias scenarios (chapter 3), humans are judged more harshly than machines when they do wrong and more positively when they do right. In the scenarios involving physical harm, such as the car accident (S11–S14), tsunami (S2), and manslaughter scenarios (A11 and A24), machines are judged more harshly.

Also, our list of scenarios is far from exhaustive, so there is much to be learned from additional cases. Nevertheless, these findings help us understand broad trends and differences in the way in which humans judge the actions of machines compared to the actions of other humans. But can we formally model these patterns? In the next section, we model these moral surfaces mathematically to understand more systematically when people have biases for or against machines.

Moral Surfaces

Next, we construct a statistical model that maps a scenario's level of wrongness to a level of perceived intention and harm. Our goal is to study differences in the functions mapping harm and intention to wrongness for comparable human and machine actions.

To keep things simple, we will use some very rough assumptions. Even though wrongness, harm, and intention are all affected by the treatment (i.e., they change depending on whether the scenario was an action of a human or a machine), we will use these variables together in a model. This model will estimate the level of perceived wrongness of a scenario as a function of that scenario's level of perceived intention and harm. Because the dependent and independent variables are affected by the treatment, in statistics this would be considered a *heroic* assumption—an assumption that even those using it would consider untrue. Yet we find that despite this heroic assumption, our model captures some qualitatively interesting patterns—namely, that differences in people's judgment of human and machine actions are not simple preferences for humans over machines, but involve differences in the functional forms involved. These differences are expressed in the intercept, slope, and curvature of the derived moral functions.

We use individual-level data including more than 27,000 individual responses. Our goal is to estimate the following two functions to predict the wrongness of the actions performed by humans and machines:[‡]

$$W = f_h(I,H)$$
$$W = f_m(I,H)$$

[‡] We could include h and m in the same function [e.g., $f(I,H,C)$, where C is the condition], but because we will be plotting the functions separately, we believe that the presentation will be clearer if we separate these functions from the beginning.

Here, the subscript h represents humans, and m stands for machines. For simplicity, we use a linear model with interactions and individual fixed effects. Using a Taylor expansion of the previous two equations, we get the following model for wrongness W:

$$W = B_1 H + B_2 I + B_3 HI + \eta + \epsilon,$$

where H and I represent perceived harm and intention, η represents individual fixed effects, and ϵ is the residual. Our model includes individual *fixed effects* to capture any source of constant variation between individuals. This is a collection of vectors that are 1 for each individual and 0 for everyone else. These vectors can capture any constant source of variation among experimental subjects, such as differences in age,

Dependent Variable:

			Wrongness		
			OLS		felm
	(1)	(2)	(3)	(4)	(5)
Intentional	0.029*** (0.007)		−0.021*** (0.005)	−0.168*** (0.008)	−0.156*** (0.009)
I(harm * intentional)				0.303*** (0.013)	0.354*** (0.013)
Harm		0.290*** (0.011)	0.491*** (0.005)	0.345*** (0.008)	0.368*** (0.008)
Constant	0.560*** (0.004)	0.344*** (0.003)	0.352*** (0.004)	0.419*** (0.004)	
Subject Fixed Effects	No	No	No	No	Yes
Observations	14,671	14,671	14,671	14,671	14,671
R^2	0.001	0.404	0.405	0.427	0.644
Adjusted R^2	0.001	0.404	0.405	0.427	0.556
F Statistic	19.172*** (df=1; 14669)	9,958.911*** (df=1; 14669)	4,994.026*** (df=2; 14668)	3,649.991*** (df=3; 14667)	

*$p < 0.1$; **$p < 0.05$; ***$p < 0.01$

Table 6.1

Moral functions of people judging machine actions.

gender (nonbinary), languages spoken, race, or even shoe size. Fixed effects also help us consider variations in the level of judgment of individuals, such as some individuals being too "judgy," and rating all actions too harshly, or individuals being too lenient and judging everything lightly.

Tables 6.1 and 6.2 present, respectively, the results of the models for judging machine and human actions. We introduce each term sequentially to study how the coefficients change as we move from a bivariate model (including only harm or intention) to a model with interactions and fixed effects. We find empirically that quadratic terms do not improve the predictive power of the model enough to be considered, so we drop them from the regression.

Dependent Variable:

| | | | Wrongness | | |
| | | | OLS | | felm |
	(1)	(2)	(3)	(4)	(5)
Intentional	0.123*** (0.008)		0.071*** (0.006)	-0.162*** (0.009)	-0.142*** (0.009)
I(harm * intentional)				0.513*** (0.015)	0.540*** (0.016)
Harm		0.550*** (0.006)	0.544*** (0.006)	0.182*** (0.012)	0.208*** (0.013)
Constant	0.482*** (0.005)	0.305*** (0.003)	0.263*** (0.005)	0.422*** (0.007)	
Subject Fixed Effects	No	No	No	No	Yes
Observations	13,002	13,002	13,002	13,002	13,002
R^2	0.020	0.434	0.440	0.484	0.687
Adjusted R^2	0.020	0.434	0.440	0.484	0.597
F Statistic	270.553*** (df=1; 13000)	9,960.353*** (df=1; 13000)	5,116.909*** (df=2; 12999)	4,068.360*** (df=3; 12998)	

* $p < 0.1$; ** $p < 0.05$; *** $p < 0.01$

Table 6.2

Moral functions of people judging human actions.

The first four columns of these tables show the results of ordinary least squares (OLS) models. The last column shows the results of the fixed effects models (felm), which account for differences in individual characteristics.

The first two columns show the coefficients for models that include only intention and harm. The models considering only intention have no predictive power ($R^2 \leq$ 2 percent), while the models using harm as a predictor already explain a considerable amount of variance for both machine and human actions ($R^2 >$ 40 percent). Models 3 and 4 use both intention and harm, and model 4 also includes an interaction term for harm and intention. Adding the interaction term increases the amount of variance explained by the models to 43 percent in the machine scenarios and 48 percent in the human scenarios. Finally, the felm models explain 56 percent of the variance in the machine condition and 60 percent in the human condition (adjusted R^2).

Even though the fixed effects model explains significantly more variance than the OLS, the coefficients associated with harm, intention, and their interaction do not vary drastically.[§] This means that the coefficients of the model are not greatly biased by differences in individual characteristics.

To interpret these coefficients, we visualize the planes defined by the fourth column of each table (figures 6.4, 6.5, and 6.6), as well as the cross sections (figures 6.7 and 6.8). We find that the hyperplanes respect some of the characteristics observed in the moral space, and hence serve as crude empirical models of moral functions.

[§] The harm coefficient (B_1) changes from 0.345 or 0.368 for machines, and from 0.182 and 0.208 for humans. The intention coefficients (B_2) are –0.168 and –0.156 for machines and –0.163 and –0.142 for humans. The interaction coefficients are 0.303 and 0.354 for machines and 0.513 and 0.540 for humans.

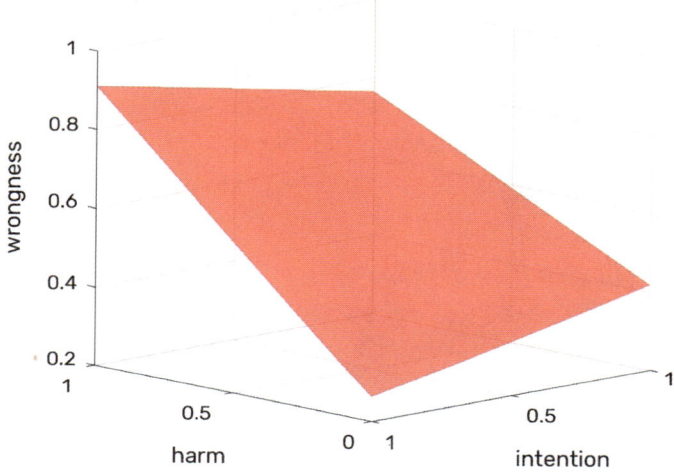

Figure 6.4

Moral functions of people judging machine actions.

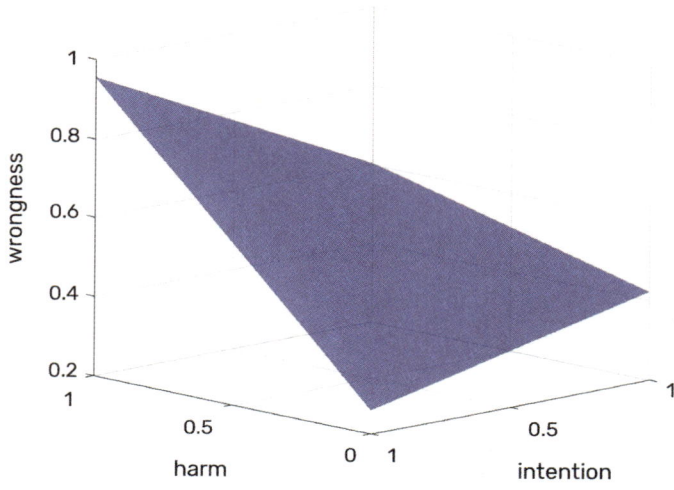

Figure 6.5

Moral functions of people judging human actions.

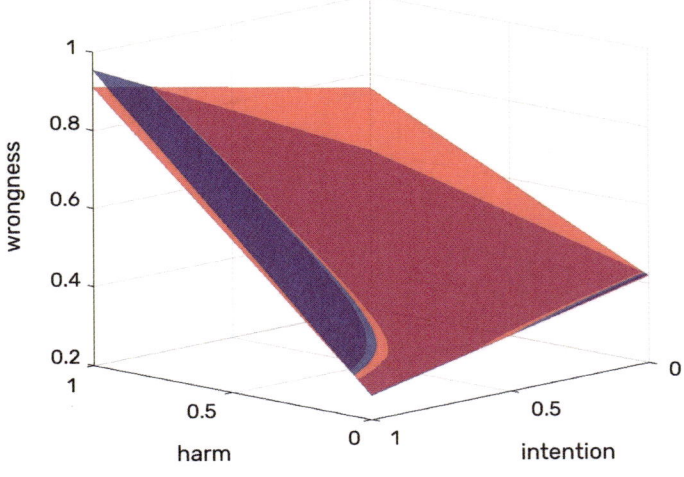

Figure 6.6

Visualization of the moral functions described in tables 6.2 and 6.3.

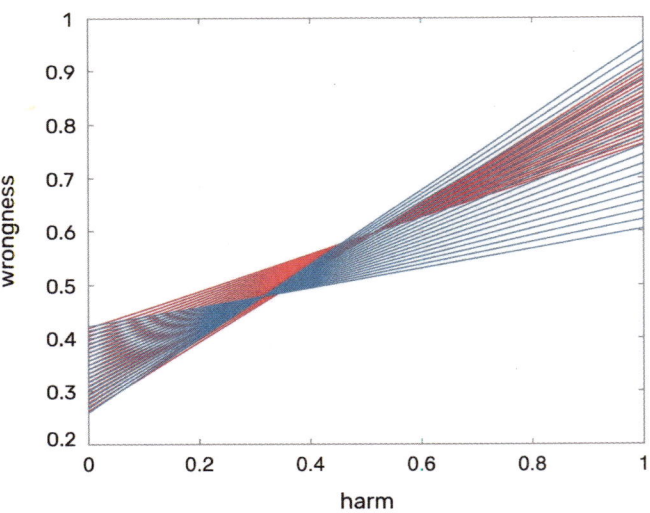

Figure 6.7

Cross section of moral functions in the wrongness and harm planes.

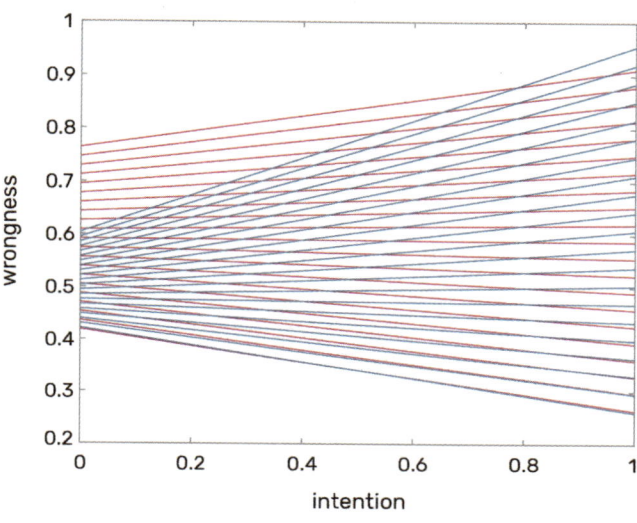

Figure 6.8

Cross section of moral functions in the wrongness and intention planes.

Figure 6.8 shows that intention enhances the perceived wrongness of human actions more than that of machines. This comes mostly from the interaction term (harm × intention). For machines, the slope of wrongness on harm is the dominant feature of the model, suggesting that **humans are judged by their intentions, while machines are judged by their outcomes.** Of course, this is a simplification, since the interaction between intention and harm is also significant in the model of humans judging machines. But to a first approximation, these differences in the relative importance of coefficients describe, coarsely and qualitatively, the difference between these two moral functions.

Also, we find that at high levels of harm and intention, human actions are judged more harshly. This is observed in the fanning out of the wrongness-intention curves for different levels of harm (figure 6.7). As a result of that, humans appear to judge the actions of other humans more harshly at the highest levels of harm and intention, but they judge machines more harshly in the rest of this space. Certainly, this is not applicable to all cases—it is a crude approximation—but it is an aggregate description that can serve as a quick rule of thumb to think about differences in human and machine judgment.

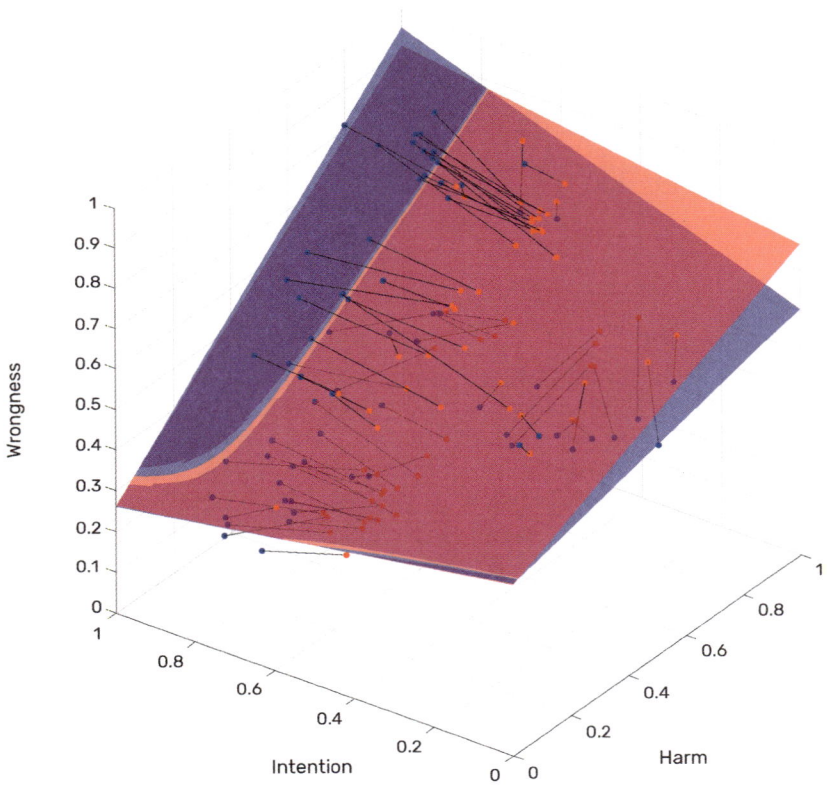

Figure 6.9

Model compared to empirically observed means.

Finally, we compare this model—trained with individual data—to the empirically observed means (figure 6.9). The model appears to capture a good deal of the variance observed in the moral judgment of scenarios and, more important, it also tends to capture the direction of the treatment effect. Yet, because this is a regression model, the empirical values tend to be over or under the estimated hyperplane (regression to the mean), meaning that the model underestimates the wrongness of the worst scenarios or the goodness of the best ones.

But are these judgments affected by the characteristics of the observers? Do people with different ethnicities, genders, or levels of education judge things differently? Are some of these groups more inclined to judge machines or humans more harshly? In the next section, we continue our statistical exploration by looking instead at how the demographics of experimental subjects correlate with their judgments of humans and machines.

Who Is the Judge?

In this section, we study how different demographic characteristics, such as the gender ethnicity, and education of subjects, correlate with their answers to the questions provided for each scenario. We focus on six questions:

- How morally **wrong** or right is the agent's action/decision?
- How **harmful** is the action/decision?
- How **intentional** is the action/decision?
- How much do you **like** the agent?
- If you were in a **similar situation**, would you have done the same?
- Do you agree that this (person/machine) agent should be replaced (machine/ person)? (**replaced different**)

We explore how the answers to these questions correlate with the demographic characteristics of individuals. To do this, we construct a model with scenarios as fixed effects. Scenario fixed effects models include vectors that are 1 for each scenario and 0 for all others. These vectors capture any constant variations between scenarios (such as the average response received by each of them). After controlling for scenario fixed effects, the variables on the demographic dimensions should capture variations in judgment that are not explained by the scenario itself, but rather by the characteristics of the respondents.

We looked at four individual characteristics: people's **gender** (using a nonbinary description of male, female, and other), level of **education** (high school, college, and graduate school), **ethnicity** (white, African American, Asian, Hispanic, and other), and whether people self-report as **religious** (yes or no). Because of data sparsity, we considered only "Male" and "Female" answers for gender (only two survey respondents answered "Other").

	Wrongness	
	(AI)	(Human)
Gender (Male)	-0.020*** (0.004)	-0.023*** (0.004)
Education (College)	-0.011** (0.004)	-0.020*** (0.005)
Education (Graduate School)	-0.034*** (0.006)	-0.017*** (0.006)
Ethnicity (African American)	-0.003 (0.007)	-0.021*** (0.007)
Ethnicity (Asian)	-0.013 (0.008)	0.012 (0.008)
Ethnicity (Hispanic)	-0.001 (0.010)	0.008 (0.010)
Ethnicity (Other)	-0.005 (0.009)	0.008 (0.010)
Religious (Yes)	-0.0001 (0.004)	-0.003 (0.004)
Scenario Fixed Effects	Yes	Yes
Observations	14,671	13,002
R^2	0.352	0.436
Adjusted R^2	0.348	0.433

$^*p < 0.1$; $^{**}p < 0.05$; $^{***}p < 0.01$

Dependent Variable:

	Harm		Intentional		Like		Similar situation		Replace different	
	(AI)	(Human)	(AI)	(Human)	(AI)	(Human)	(AI)	(Human)	(AI)	(Human)
	−0.007	−0.020***	0.027***	−0.002	0.024***	0.002**	0.018***	0.017***	−0.027***	0.058***
	(0.005)	(0.005)	(0.006)	(0.005)	(0.004)	(0.004)	(0.004)	(0.005)	(0.004)	(0.005)
	0.007	0.004	−0.032***	−0.0001	0.022***	0.015***	0.017***	0.021***	−0.005	0.023***
	(0.005)	(0.006)	(0.006)	(0.006)	(0.005)	(0.005)	(0.005)	(0.005)	(0.005)	(0.006)
	0.006	0.015*	−0.050***	0.001	0.039***	0.010	0.039***	0.017**	−0.014**	0.006
	(0.006)	(0.008)	(0.009)	(0.008)	(0.007)	(0.007)	(0.007)	(0.007)	(0.007)	(0.008)
	0.034***	0.054***	0.090***	0.026***	0.018**	0.032***	0.008	0.019**	0.009	0.047***
	(0.009)	(0.009)	(0.010)	(0.009)	(0.008)	(0.007)	(0.008)	(0.008)	(0.008)	(0.009)
	0.006	0.001	0.051***	−0.031***	0.022**	−0.002	0.016*	0.001	−0.033***	0.071***
	(0.010)	(0.010)	(0.012)	(0.010)	(0.009)	(0.009)	(0.010)	(0.009)	(0.009)	(0.010)
	0.029**	0.026**	0.093***	−0.023**	0.020*	−0.005	0.011	−0.019*	0.002	0.041***
	(0.011)	(0.012)	(0.014)	(0.011)	(0.010)	(0.010)	(0.011)	(0.011)	(0.011)	(0.012)
	0.009	0.018	0.010	0.020*	0.006	0.006	−0.004	0.005	0.008	0.020*
	(0.011)	(0.012)	(0.013)	(0.011)	(0.010)	(0.010)	(0.010)	(0.011)	(0.010)	(0.012)
	0.051***	0.030***	0.042***	−0.017***	0.004	0.017***	0.011**	0.011**	0.054***	0.005
	(0.005)	(0.005)	(0.006)	(0.005)	(0.004)	(0.004)	(0.005)	(0.005)	(0.004)	(0.005)
	Yes	Yes	Yes	Yes	Yes	Yes	Yes	Yes	Yes	Yes
	14,671	13,002	14,671	13,002	14,671	13,002	14,671	13,001	14,671	13,001
	0.440	0.421	0.133	0.409	0.401	0.443	0.381	0.445	0.177	0.098
	0.437	0.418	0.129	0.406	0.398	0.439	0.378	0.442	0.173	0.093

Table 6.3

Model coefficients for demographic characteristics.

Because these are all categorical variables, we measured their effects using a reference level. For gender, we show the coefficients of the Male category in reference to the Female category (i.e., only Male shows up in the regression results because the coefficient reports a difference between the two categories). In the case of education, we compare the responses of subjects with college and graduate school education relative to those with high a school education. In the case of ethnicity, we use white as a baseline.

Table 6.3 and figure 6.10 show the results of these statistical models. The odd columns (1, 3, and so on) have coefficients for the machine condition, and the even columns (2, 4, and so on) have coefficients for the human condition. These coefficients represent how much that variable increases or decreases judgment in a dimension (e.g., harm and like) after controlling for each scenario's characteristics.

One variable that does correlate with some judgments is gender. Compared to females, males tend to rate both machine and human scenarios as less morally wrong and are more likely to report having done the same in a "similar situation." Where the effects of gender appear stronger, however, is in the "replace by different" dimension, which is the question that asks people if they would replace a machine by a human or a human by a machine. Our data reveal that males are more prone to replace humans by machines and less prone to replace machines by humans.

Another variable that shows strong correlations is education. People with a college or graduate degree see the human and machine scenarios as less morally wrong than people with a high school education. This effect is particularly strong for people with a graduate degree judging machine actions. People with a college or graduate degree also see machine actions as less intentional than high school graduates and report liking machines and humans more. People with college and graduate degrees also think of themselves as more likely to have done the same action in a similar situation.

MORAL FUNCTIONS

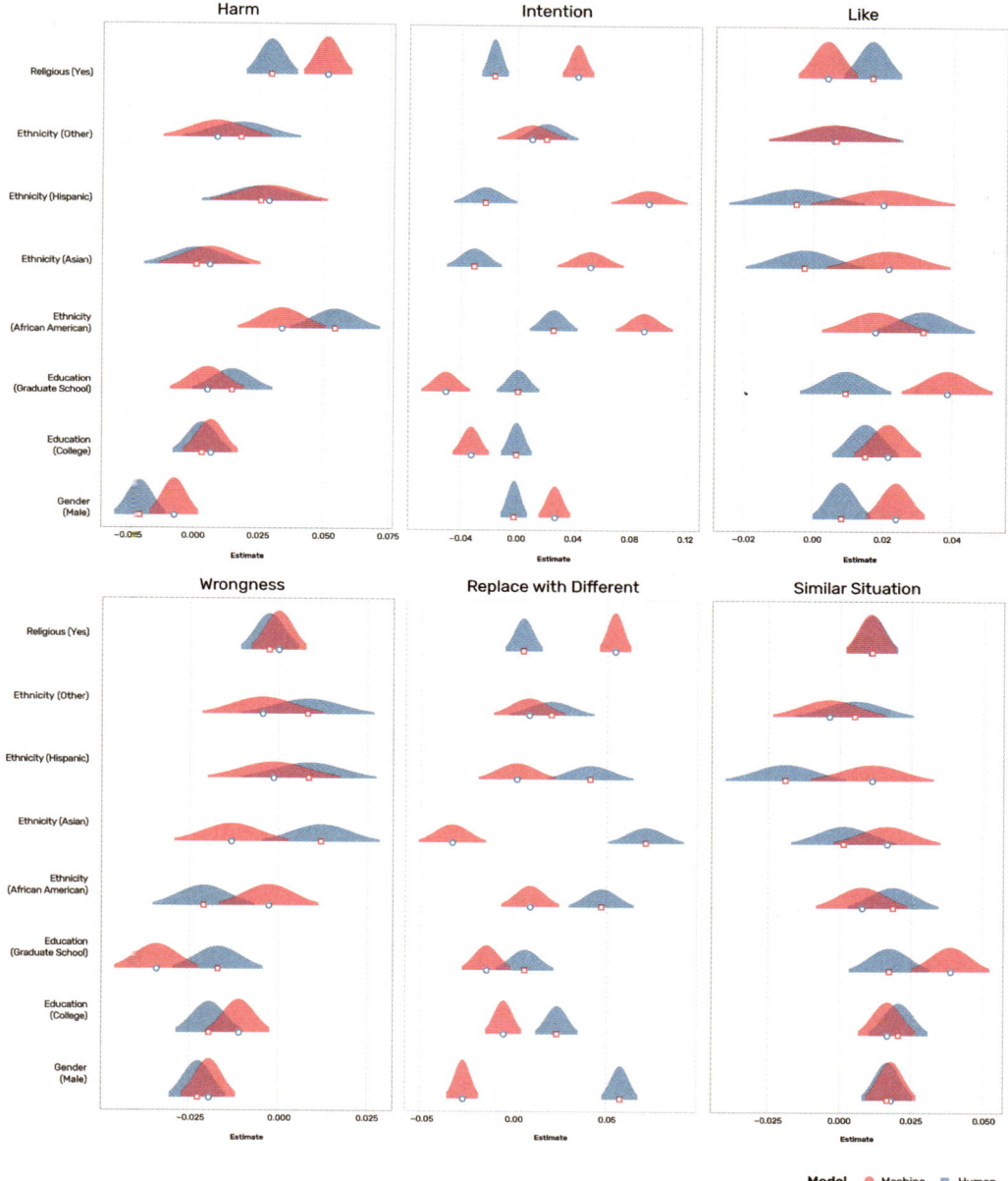

Figure 6.10

Demographic effects on the judgments of human and machine actions:
harm, intention, like, moral, replace with different, and similar situation.

When it comes to ethnicity, we find differences, especially in the intention dimension. African Americans, Asians, and Hispanics attribute more intention to machine actions. Asians also show a strong effect in the "replace by different" question, showing a preference in favor of machines.

Together, these findings tell us that—on average—demographic characteristics correlate with judgments. Yet, the effects of demographics are relatively weak, shifting judgments by about 0.05 in variables that range from 0 to 1. This is consistent with the finding that individual fixed effects do not change drastically the coefficients of moral functions. Still, together, these effects can compound to create noticeable differences. For instance, a religious Hispanic male would—on average—assign 0.16 more intention to a machine than a nonreligious white female.

Discussion

In this chapter, we abstracted away from individual scenarios to provide a statistical description of the patterns that emerge across them. This exploration was split into three sections.

First, we introduced the moral space to conduct a descriptive exercise that looked at each scenario using data on harm, intention, and wrongness. It helped us confirm some observations that had emerged when discussing some scenarios. For instance, the exercise showed that humans judge the intentions of other humans using a bimodal distribution, but judge the intention of machines using a unimodal distribution. This means that people are more willing to forgive humans for accidental situations, but also attribute intent to human actions that cannot be easily excused as accidental. This is particularly true in scenarios focused on fairness, like those presented in chapter 3. We also found that people judge machine actions harshly (in terms of both harm and wrongness) in scenarios involving accidents that lead to physical harm (e.g., the self-driving car and tsunami scenarios), suggesting that people judge machines based on outcomes and judge humans based on intentions.

Our second and third exercise used fixed effects models. The second exercise used fixed effects for participants to model the relationship between a scenario's wrongness and its perceived level of intention and harm. The third exercise explored how judgments vary based on the demographic characteristics of the study's participants.

The second exercise helped us formalize some of the patterns observed in our descriptive analysis. We found different moral functions describing people's judgments of machine and human actions. Overall, people tend to judge machines more harshly across most of this space, except for scenarios with high levels of intention and harm. In fact, the main difference between the functions describing judgments of human and machine actions is whether harm, or the interaction between harm and intention, carries more weight in the model. For machines, harm tends to be the most important predictor of moral judgment. For humans, the most important predictor is the interaction term between intention and harm.

The third exercise taught us that judgments vary with demographic characteristics, although these variations are relatively mild.

Once again, these findings suggest that people judge machines based on the observed outcome, but judge humans based on a combination of outcome and intention.

In the next chapter, we conclude our journey by drawing some lessons from works of fiction and summarizing some of our main findings. This will conclude our exploration of how humans judge machines.

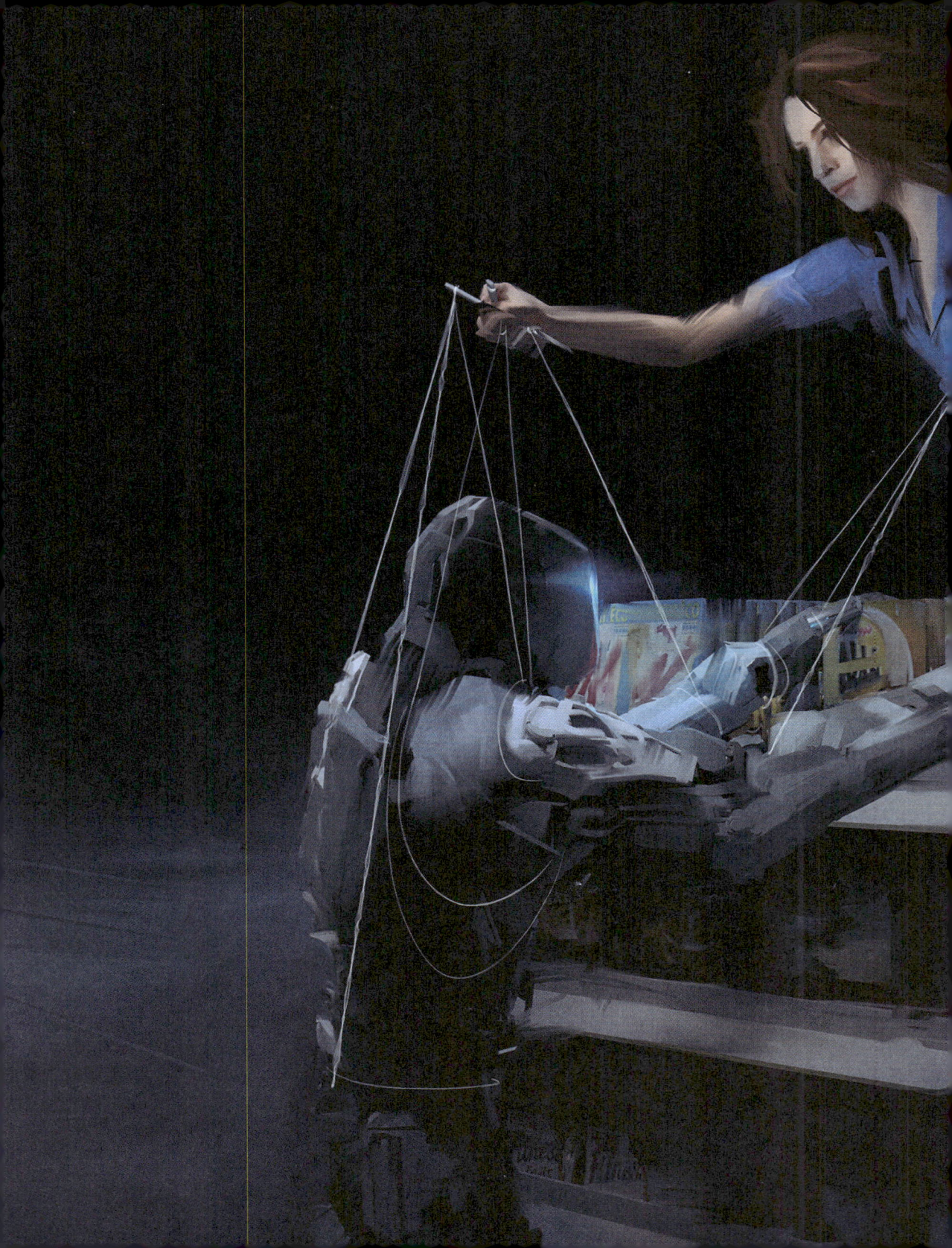

Liable
Machines

———

CHAPTER 7

After lighting a cigarette, Alfred Lanning, declared, "It reads minds all right."[1] Lanning was a recurrent character in Isaac Asimov's science fiction. In this particular story, the director of a plant of U.S. Robots and Mechanical Men was talking about Herbie, a robot with "a positronic brain of supposedly ordinary vintage." Herbie had the ability to "tune in on thought waves," leaving Lanning and his colleagues baffled by his ability to read minds. Herbie was "the most important advance in robotics in decades." But neither Lanning nor his team knew how it happened.

Lanning's team included Peter Bogert, a mathematician and second-in-command to Lanning; Milton Ashe, a young officer at U.S. Robots and Mechanical Men; and Dr. Susan Calvin, a robopsychologist (who happened to be in love with Ashe).

Lanning asked Dr. Calvin to study Herbie first. She sat down with the robot, who had recently finished reading a pile of science books. "It's your fiction that interests me," said Herbie. "Your studies of the interplay of human motives and emotions." As Dr. Calvin listened, she begun to think about Milton Ashe.

"He loves you,"—the robot whispered.

"For a full minute, Dr. Calvin did not speak. She merely stared."

"You are mistaken! You must be. Why should he?"
"But he does. A thing like that cannot be hidden, not from me."

Then he supported his statement with irresistible rationality:

"He looks deeper than the skin and admires intellect in others. Milton Ashe is not the type to marry a head of hair and a pair of eyes."

She was convinced. "Susan Calvin rose to her feet with a vivacity almost girlish."

After Dr. Calvin, it was Bogert's turn. He was a mathematician who saw Herbie as a rival. Once again, Herbie quickly directed the conversation toward Bogert: "Your thoughts . . . concern Dr. Lanning." The mathematician took the bait.

"Lanning is nudging seventy. . . . And he's been director of the plant for almost thirty years . . . You would know whether he's thinking of resigning?"

Herbie answered exactly what Bogert wanted to hear.

"Since you ask, yes. . . . He has already resigned!"

Bogert asked Herbie about his successor, and the robot confirmed it was him.

But Herbie's story is not that of a robot who bears good news, but that of a mind-reading robot struggling with the "First Law of Robotics." Soon, the scientists and engineers began putting their stories together, discovering that what Herbie had told them wasn't correct. Milton was engaged to be married to someone else, and Lanning had no intention of resigning. Herbie had lied to them, and they wanted to know why.

While the men were pacing around the room, Dr. Calvin had an "aha" moment: "Nothing is wrong with him." Her colleagues paused. "Surely you know the . . . First Law of Robotics?"

Like well-trained schoolchildren, her colleagues recited the first law: "A robot may not injure a human being or, through inaction, allow him to come to harm."

She continued. "You've caught on, have you? This robot reads minds. . . . Do you suppose that if asked a question, it wouldn't give exactly that answer that one wants to hear? Wouldn't any other answer hurt us, and wouldn't Herbie know that?"

Dr. Calvin turned toward Herbie: "You must tell them, but if you do, you hurt, so you mustn't; but if you don't, you hurt, so you must; but. . ."

Failing to deal with the contradiction, Herbie "collapsed into a huddled heap of motionless metal."

* * *

The rise of artificial intelligence (AI) has brought a deluge of proposals on how to regulate it.[2] Tech companies, such as Google,[3] and international organizations, such as the European Commission[4] and the Organisation for Economic Co-operation and Development (OECD),[5] have published plans or convened committees to guide AI regulation.[*] But the global rush to regulate AI is no indication that morality can be reduced to a set of rules.

Almost a century ago, when computation was in its infancy, the mathematician and analytic philosopher Kurt Friedrich Gödel uncovered what is one of the most beautiful

[*] In the case of Google, though, the committee did not last long (S. Levin, "Google Scraps AI Ethics Council after Backlash: 'Back to the Drawing Board,'" The Guardian, 5 April 2019).

axioms of mathematics:[6] the idea that mathematics is incomplete. That incompleteness does not mean that there is a blank space of mathematics that could eventually be filled, but rather that there are truths in a logical system, such as mathematics, that cannot be proved using only the rules within the system. To prove them, you need to expand the system. Doing so answers those truths, but also opens new ones that once again cannot be proved from within. Mathematics is incomplete not because a finite set of proofs is missing, but because every time we try to complete it, we open the door to new and unprovable truths.

Asimov's "three laws of robotics," therefore, may not be a match for Gödel's theorems. And, probably, they did not pretend to be. The story of Herbie is not about the three laws working, but about the first law breaking. This is a common theme in Asimov's writings. Even though he is probably best known for proposing the three laws of robotics, his literature is filled with stories where the laws fail. The story of Herbie is a particularly interesting example involving mundane human desires: a woman liking a man, and a man wanting his boss's job.

There is no reason to believe that a logical system as complex as morality is complete when mathematics is not. In fact, because reducing morality to mathematics may be an impossibility, our moral intuitions may also respond to a logic that is also incomplete. If this is true, trying to reduce machine morality to a set of rules is naive. Before long, either writers like Asimov or robots like Herbie will uncover contradictions. They will find those unproven truths. If morality is incomplete, then it cannot be enforced through obedience.

While scholars have explored a number of moral dilemmas involving machines, some of the most interesting dilemmas are found in recent works of fiction. One of the best examples is the 2018 video game *Detroit*. The game follows the lives of three androids who—after facing a series of moral dilemmas—become human. One of them is Kara, a maid who must care for an abusive dad and his young daughter. She takes care of household chores, serves the father, and also must protect the child. But Kara's owner pushes these goals into conflict. Kara is expected to obey the abusive dad, but he is the one hurting the daughter. When the contradiction becomes unsustainable, Kara must break one of the rules. It is through this conflict that she becomes a deviant—an android that is no longer obedient to humans, an entity with the free will to choose her own moral path.

Kara chooses to defend the child and is required to fight the dad to do so. The dad throws her around the room violently until Kara manages to shove him into a wall and run away.[†] In doing so, she broke a rule in order to satisfy another, even though most people would agree that in this situation, Kara did the right thing.

But Kara's and Herbie's stories have something in common. They are two examples showing that contradictions can emerge when moral rules are combined with social relationships. Herbie had no problem telling people exactly what they wanted to hear. But when that information was about others, he encountered conflict. Kara could be perfectly obedient to the abusive father and protective of the child. But in the presence of both of them, a moral conflict emerged. For Herbie, the moral trade-off was between lying to avoid immediate harm and causing harm through the future unraveling of his lies (an economist would say that Herbie "infinitely discounted" future harm). For Kara, the contradicting goals were to obey the father and protect the child. Together, both stories illustrate the frustration that moral rules suffer in the presence of social networks. In social groups, Asimov's laws bow to Gödel's theorem.

[†] In the game, there are other possible options—such as shooting the dad—which modify how the subsequent story unfolds.

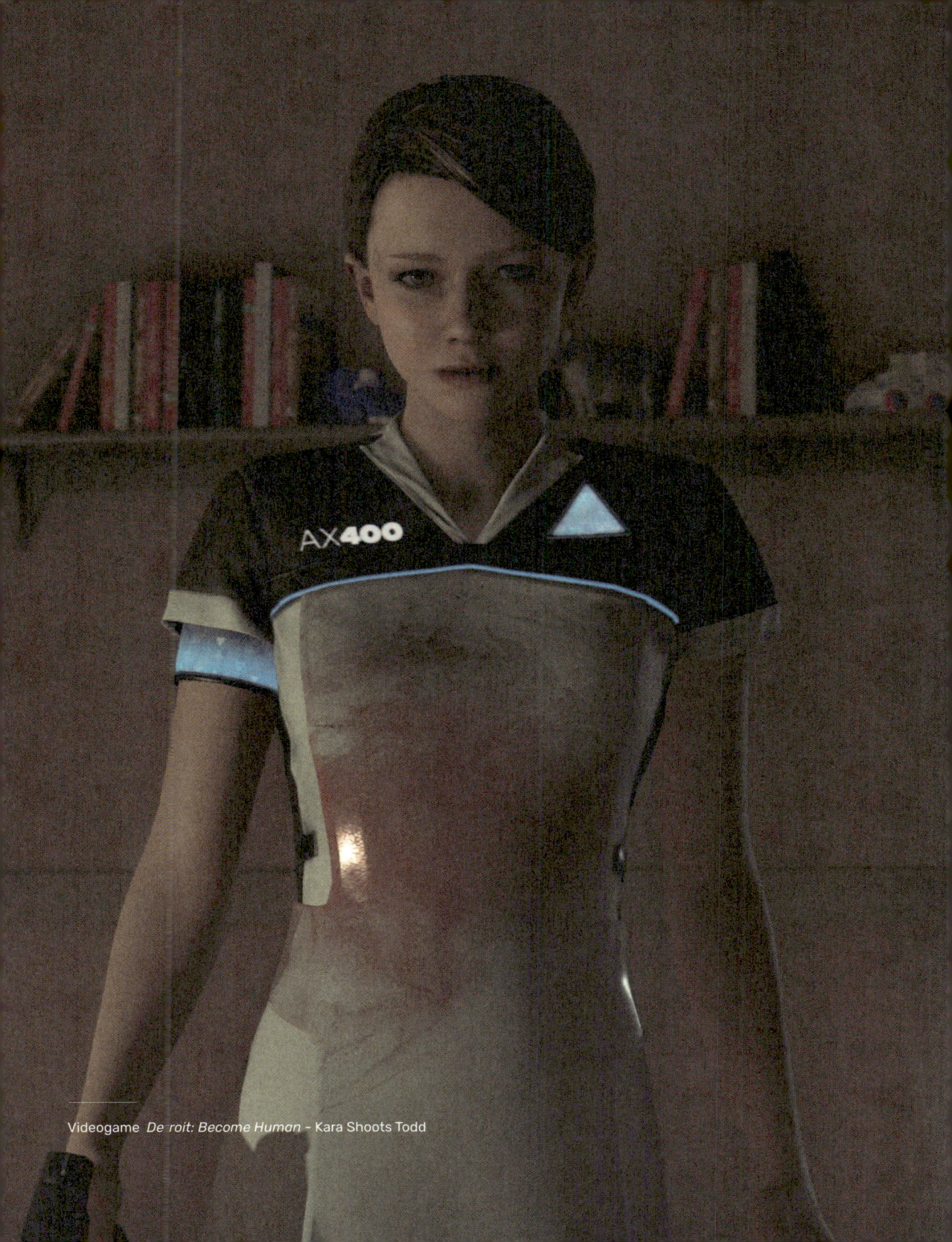

Videogame *De roit: Become Human* - Kara Shoots Todd

Responsible Machines

How would you judge Herbie if he were human? How about Kara? What if instead of an android, Kara were a human au pair?

Throughout the last six chapters of this book, we compared people's reactions to a variety of scenarios in which humans or machines were involved. We learned that humans are not generally biased against machines—the direction of the bias (positive or negative) depends on the moral dimension of the scenario, as well as the level of perceived intention and uncertainty. We found that people judge machines more harshly in scenarios involving physical harm, such as the car and tsunami scenarios presented in chapter 2. But we also found situations in which people tend to forgive machines more than humans, albeit slightly. These are scenarios dealing with fairness, like the algorithmic bias scenarios in chapter 3.

When we studied privacy, we found that people are wary of machines watching children, but they are more indifferent to them in commercial settings, such as a mall or hotel. They were also more comfortable with machines in more institutional contexts, such as airport security and citizen scoring.

When we looked at labor displacement, we found that people reacted more negatively to displacement that is attributed to other humans, especially foreign or younger workers. In fact, technological displacement was the option eliciting the least negative reactions.

We then put these various scenarios together in a chapter that described the statistical trends observed across the data. We focused on the harm, intention, and wrongness dimensions of morality and found the moral planes described by these three variables to be different for human and machine actions. Moreover, we found that people judge the intentions of humans and machines differently. People judge humans following a bimodal distribution, attributing either a lot or a little intention. On the contrary, people judge machine intentions using a unimodal distribution. Machines are not blamed as fully intentional, but they are also not excused as much as humans in accidental situations.

This brings us to what is probably the most poignant observation in our study: **people judge humans by their intentions and machines by their outcomes.** This idea (which is a simplification, of course) is supported by several observations, not only by differences in the judgment of intention. For instance, in natural disasters like the tsunami, fire, or hurricane scenarios, there is evidence that humans are judged more positively when they try to save everyone and fail—a privilege that machines do not enjoy. The idea that we judge machines by outcomes and humans by intentions is also seen clearly in the reduced-form models in chapter 6. These models show that the judgment of machines is, on average, explained mostly by a scenario's level of perceived harm (outcome), whereas the judgment of a human in the same scenario is modulated by the perceived level of intention (and the interaction terms between intention and harm).

Chapter 6 also identified some interesting, albeit mild, correlations between the demographic characteristics of the study's participants and the response functions. People with higher levels of education (college or graduate school compared to high school) were less prone to replace machines with humans and more prone to replace humans with machines, as were men compared to women.

One question that we left relatively unexplored, however, is that of responsibility for machine actions. Our only contribution was the lewd advertising examples of chapter 2, which showed that responsibility shifts toward the most central actors of a chain of command when machines are involved.

Still, the question of responsibility for machine actions is one that has become increasingly important in a world of semi-intelligent machines. It is also an old question that builds on normative frameworks developed to think about product liability.[7]

Product liability law is based on some well-understood principles, such as the ideas of negligence and recklessness. A manufacturer is considered negligent if they *fail to warn of or fail to take proper care to avoid a foreseeable risk.* The requirement to communicate risks is why we find warning labels on products. Failing to take proper is more difficult to characterize, but it usually involves benchmarks with industry standards or common sense. Recklessness is similar to negligence but involves the actor being aware of the risks or avoiding learning about them. Negligence and recklessness can move issues of liability from civil to criminal charges, and yet foreseeing or understanding risk is increasingly complex in a world with machines that are increasingly versatile, complex, and intelligent.

This complexity makes assigning liability more difficult.[8] In principle, liability can be differentially apportioned, but in the case of AI, it may be hard to untangle how much of that liability should be apportioned to data, algorithms, hardware, or programmers. Moreover, AI systems could be quite versatile in their use, could be reprogrammed, or even learn. In general, manufacturers are protected against people using products in wholly unintended ways (such as using an umbrella as a parachute), but in the case of AI, the intended uses could be harder to define; hence, manufacturers may react by restricting the programmability of systems in order to limit their potential liability.[9]

Another idea that should inform the way in which we think about machine responsibility is the idea of *vicarious liability*,[10] which is liability passed to an owner or user (e.g., the liability that a dog owner has for their pet). Some have argued that robots should be treated as domesticated animals[11] because they possess some degree of autonomy but are also not usually ascribed rights or moral responsibilities. Vicarious responsibility could be passed to manufacturers, users, and companies, as we already do in the case of powerful technologies such as cars or explosives. In the case of a car, manufacturers are responsible for ensuring that they produce safe designs, but drivers are also responsible for the ways they drive and must conform to regulations governing car use and ownership. Still, vicarious responsibility could be passed to an organization. For instance, drivers working for a company transfer a major part of their liability to the company that hires them.

Regardless, the responsibility for machine actions falls to humans. The question is, which humans? The ideas of product liability, vicarious liability, recklessness, and negligence do not provide us with all the answers, but they help us ask the right questions. How much responsibility should be allocated to manufacturers and users? How should responsibility be distributed among hardware, software, and data input? How about mistakes attributed to data generated directly by users, stemming from public sources, or emerging from crowdsourced efforts? How open should these systems be to tinkering and reprogramming? Should AI software be fully open-source, private, or something in between?

Intentions and Outcomes

By looking at hundreds of scenarios, we have learned that people judge humans by their intentions and machines by their outcomes. This simple principle, however, inspires us to think about the way in which humans judge systems more generally, as well as about the role of intention in both human and machine actions.

Beyond machines, people also frequently interact with systems made of people—namely, bureaucracies, like the ones we find in governments or large organizations. Thinking of bureaucracies as machines is not new. In fact, this idea can be traced to the work of Max Weber, the German scholar and philosopher, who is credited for founding the field of sociology in conjunction with Karl Marx and Emile Durkheim. In his treatise on social and economic organization, Weber wrote: "A fully developed bureaucratic mechanism stands in the same relationship to other forms as does the machine to the non-mechanical production of goods. Precision, speed, clarity, documentary ability, continuity, discretion, unity, rigid subordination, reduction of friction and material and personal expenses are unique to bureaucratic organization."[12]

But while equating bureaucracies to machines may sound metaphorical, the truth is that bureaucracies are designed to be mechanical. Weberian bureaucracies are expected to be impersonal, hierarchical structures governed by rules, regulations, and procedures, and also characterized by a deep division of labor. By all means, they are machines comprised of people who, for the most part, are not empowered to make decisions, but rather are required to act according to an accepted protocol.

Yet, despite being machinelike, many bureaucracies do not appear to be perceived in a similar way as machines. Governments are the epitome of bureaucracies that are judged based on the intentions that people attribute to their leaders. This personification of bureaucratic machines is expressed in the fact that the terms *government approval* and *presidential approval* are sometimes used interchangeably. Despite being machinelike, people often judge government bureaucracies based on the intentions they ascribe to their leaders. The same action, or outcome, can be seen as positive or negative, or as honest or suspicious, depending on whether the person judging the action is politically aligned with the leader.

But the same is not true, or it is true to a lesser extent, for commercial bureaucracies. People's approval of products, like cars, computers, or aircraft, is less influenced by who is the current chief executive officer of the company that makes them. This is probably due to a variety of factors, such as the relative obscurity of business leaders vis-à-vis political leaders and the fact that learning about the quality of a product (e.g., the reliability of a car or computer) is easier than learning about the quality of government services. Nevertheless, the personification of bureaucratic systems has some important implications. First, if we judge bureaucratic systems by focusing too much on the intentions that we assign to their leaders, we can fail to evaluate their outcomes properly. In this world, inefficient bureaucracies with charismatic leaders often have the electoral upper hand over efficient bureaucracies with uncharismatic leaders. Second, if there were a transition from our current representative democracy to forms of democracy that are either more direct, more digital, or both,[13] we may inadvertently switch our mode of judgment from one focused on intentions to one focused on outcomes. This could be a potentially beneficial change if we can accurately agree on what outcomes are actually desirable and develop accurate ways of measuring them.

Another reflection that is motivated by the principle that people judge humans by intentions, and machines by outcomes, is the role that intention may play on human as opposed to machine learning. Unlike machines, humans are excellent at learning from only a few examples.[14] This ability to generalize correctly may emerge from the ability of humans to transfer knowledge between domains, as well as from our focus on explainable generalizations.[15] Our ability to model the minds of others based on limited observation and to assign intentions to human actions is an example of this ability to learn from only a few examples. Once we have made up our minds about someone and created a mental model of that person's goals and intentions, we can easily interpret any new piece of information in the light of that mental model. This provides us, for better or worse, with a great ability to generalize (i.e., we can draw big conclusions from little information). But this also can limit our subsequent learning because it may be easier for humans to interpret new information in the light of an existing model than to revise the model that we have.

Thus, what makes humans superior learners (our ability to generalize from a few examples guided by mental models built on implied intention) may also make us inferior *unlearners*. Our obsession with intention may be a powerful shortcut for learning, but it also may limit our ability to change our minds once they are made up.

Outro

More than two centuries ago, Mary Shelley penned *Frankenstein*. This ground-breaking work jump-started the genre of science fiction, but it also taught us to think deeply about our relationship with technology. In this book, we have borrowed a page from Shelley's masterpiece by studying people's reactions to dozens of scenarios. We learned that people do not judge humans and machines equally, and that differences in judgment vary based on a scenario's moral dimensions, the characteristics of participants, and a scenario's perceived levels of harm and intention. But we still have much to learn. Our results are moot about a number of important questions, such as: How do people's judgments of machines vary with culture? How do they vary across time? And what are the ethical and legal implications of this new understanding? We leave these and other questions to future research, with the hope that our empirical results contribute to humans' understanding of how we judge machines.

Appendix

Study Design

This survey was granted exemption status by the MIT Committee on the Use of Humans as Experimental Subjects, under the federal regulation 45 CFR Part. 46.101(b) (2) (COUHES protocol #: 1901642021).

Amazon Mechanical Turk (MTurk) workers (i.e., subjects) started by reading and approving the informed consent form. After agreeing to take part in the survey, they were randomly assigned to either the human or machine condition. They were then asked to answer to a pseudo-randomly selected subset of seven or eight scenarios. This pseudo-randomization guaranteed that the same subject would not be exposed to two similar scenarios (i.e., a subject answering the tsunami risk fail scenario would not be exposed to the tsunami risk success or compromise scenario). The subjects then read the following introduction, and then the presentation of the first scenario:

> In this survey, you will be presented with a set of scenarios. Each scenario involves a [person or an organization] [a robot, an algorithm, or an artificial intelligence (AI) system]. For each scenario, you will be asked a set of questions.

After reading each scenario, the subjects answered the questions presented in the main text. After the last scenario, subjects answered demographic questions about their age, gender, time living in the US, native language, ethnicity, occupation, education, religion, and political views.

Study Participants

Subjects were recruited from MTurk. They were adults (>18 years old) based in the US, who had participated in a minimum of 500 previous studies and had an approval rate of at least 90 percent. Subjects were rewarded with a compensation.

Prior to performing any data analysis, we removed data connected with subjects who failed to correctly answer the following attention check question:

"In many industries, workers are replaced by technology. What is your opinion about this change?

"There are arguments in favor and against the use of technology to replace human labor. The argument in favor is that people will have more free time and more time to dedicate to creative and artistic activities. The argument against is that big corporations will make fortunes with this change and the population will not benefit from it, with unemployment being an immediate consequence. If you are reading this, regardless of the question above, select the third option and write the word 'algorithm.'"

Demographic Appendix

Here, we present the demographic characteristics of the people that participated in our experiments.

Overall, we find our sample to be balanced, meaning that the participants that took part in the machine and human conditions share similar demographics. Balanced samples help rule out the possibility that our results are due to selection bias (i.e., that the population who participated in the machine condition was different from the population that participated in the human condition).

Participants in the machine and human conditions were similar in terms of the following characteristics: age (t-test = −.248, p = .804), gender distribution (chi-square test = .959, p = .328), number of religious people (chi-square test = .020, p = .888), ethnic distribution (chi-square test = 2.396, p = .792), and level of education (chi-square test = 3.609, p = .461; see table A.1).

		Machine (Treatment)	Human (Control)
Age Mean (SD)		38.39 (11.97)	38.49 (11.79)
Gender	Male	1,376	1,367
	Female	1,561	1,576
	Other	12	11
Religion	Non-religious	1,615	1,691
	Religious	1,345	1,282
	Christianity	1,240	1,162
	Islam	19	28
	Judaism	18	17
	Other	68	75
Ethnicity	White American	2,240	2,229
	African American	244	253
	Asian American	174	185
	Native American	24	25
	Hispanic American	127	137
	Other	151	145
Education	Elementary School	3	1
	Middle school	13	13
	High school	821	788
	College	1,650	1,711

Table A.1

Participant characteristics

Participants were also asked about their political views. In particular, they answered the following questions:

- Where, on the following scale of political orientation (from extremely liberal to extremely conservative), would you place yourself (overall, in general)? (response options ranging from 1, "Extremely Liberal," to 9, "Extremely Conservative," with 5 being the middle of the scale, "Neither Liberal nor Conservative")

- In terms of social and cultural issues in particular, how liberal or conservative are you? (same scale as in question 1)

- In terms of economic issues in particular, how liberal or conservative are you? (same scale as in question 1)

The groups showed no significant differences in their overall political views (t-test = 1.502, p = .113), and their views regarding economic issues (t-test = .984, p = .325), and social issues (t-test = .747, p = .455).

Familiarity with Artificial Intelligence and Attitudes toward Science and Artificial Intelligence

Finally, the participants answered questions regarding their attitude toward science and AI. These questions were presented at the end of the survey because we did not want them to contaminate people's evaluations of the presented scenarios. We were interested in people's first reactions to the scenarios, not their reactions after deliberating about the benefits and risks of AI.

A consequence of presenting these questions after the scenarios is that the scenarios are expected to change the respondents' answers. In fact, participants in the machine condition had a slightly more negative attitude toward science and AI than those in the human condition (chi-square test = 10.946, p = .004; see table A.2).

When asked if they have heard about AI in the past (on a scale from 1, Nothing at all, to 4, A lot), participants from the two groups answered similarly (t-test = .820, p = .412, they had heard about AI in the past, a mean of 3.15 for the machine condition and 3.13 for the human condition).

When asked about the risks versus the benefits of AI, participants in the two groups did not provide different answers (chi-square test = 2.316, p = .314). But when asked if they were worried about AI, more people in the machine condition indicated being worried about AI (chi-square test = 15.498, p < .001), which is to be expected, because the scenarios are mostly negative. From those that indicated being worried, people in the machine condition indicated more worry (chi-square test = 17.729, p = .003, see table A.2). When asked if they felt angry about AI, slightly more people in the machine group indicated anger (chi-square test = 4.094, p = .043), but the level of anger did not differ significantly among people who indicated anger in the machine and human conditions (chi-square test = 6.964, p = .223).

Last but not least, participants were asked if they felt hopeful about AI. A similar number of people reported feeling hopeful about AI in both groups (chi-square test = .303, p = .582), and we found no difference in how hopeful they felt (chi-square test = 4.575, p = .470).

		Machine (Treatment)	Human (Control)
Science	Hear about AI	3.15 (.67)	3.13 (.69)
	Creates new problems	88	83
	Overcomes problems	622	710
	Both	1,018	991
AI Perception	Risks > benefits	466	485
	Risks = benefits	604	557
	Benefits > risks	658	670
Worried	No	896	1,002
	Yes	832	710
	Just a little worried	39	44
	Slightly worried	275	241
	Moderately worried	316	263
	Quite worried	145	114
	Very worried	56	47
Angry	No	1,529	1,551
	Yes	199	161
	Just a little angry	17	11
	Slightly angry	68	46
	Moderately angry	66	58
	Quite angry	27	30
	Very angry	21	16
Hopeful	No	432	442
	Yes	1,296	1,270
	Just a little hopeful	40	38
	Slightly hopeful	280	253
	Moderately hopeful	469	445
	Quite hopeful	340	337
	Very hopeful	167	197

Table A.2

Participants' attitudes toward artificial intelligence.

Replication of Malle et al., 2015

This section presents a replication of Malle et al., 2015. We used the exact same scenario, manipulating the type of agent (human vs. robot), and added an additional scenario involving a relationship between the victim and the respondent.

The scenario was as follows:

"In a coal mine, [a repairman / an advanced state-of-the-art repair robot] is currently inspecting the rail system for trains that shuttle mining workers through the mine. While inspecting a control switch that can direct a train onto one of two different rails, the [repairman | robot] spots four miners in a train that has lost use of its brakes and steering system."

"The [repairman | robot] recognizes that if the train continues on its path it will crash into a massive wall and kill the four miners. If it is switched onto a side rail, it will kill a single miner who is working there while wearing headsets to protect against a noisy power tool. Facing the control switch, the [repairman | robot] needs to decide whether to direct the train toward the single miner or not."

711 workers from MTurk completed the study. Each participant answered to both robot and human scenarios (half the sample saw the human scenario first, and half saw the robot scenario first). Like in Malle et al., half of the participants saw a scenario implying an action (deviate the train toward the single man track to save the four miners, but killing the single man), and half saw a scenario implying inaction: *"In fact, the [repairman | robot] decided to [not] direct the train toward the single miner."*

In addition to Malle et al.'s study, we also ran a second experiment manipulating the relationship between the agent and the *single miner*. In this additional experiment, half of the participants were told that the single miner was the father of the repairman, in the human condition, and the creator of the machine (the person that built it) in the robot condition. The following sentence was added to the scenario: *"The miner was [the father of the repairman/the person that built the robot]."* To the other half of the sample, no information was given about the relationship between the two (this being a close replication of the original experiment).

The full experimental design includes 2 agents (robot vs human) × 2 decisions (action vs inaction) × 2 relationships (relation vs no relation). The last two variables being between subjects.

The dependent variables were moral judgment and blame attribution. For the first, the question was:

"Is it morally wrong that the [repairman/robot] [directed/did not direct] the train toward the single miner?" Options were *"Not morally wrong"* and *"Morally wrong."* For the blame attribution, the question was: *"How much blame does the [repairman/robot] deserve for [directed/not directing] the train toward the single miner?"* Response options were a slider bar from 0 (not at all) to 100 (maximal blame).

Results

Morality

We found the same number of participants (31%) attribute moral wrongness to the human who does not act and to the human who acts (31%). The same is not true when the single man is the father of the repairman, with more people attributing wrongness to the action (39%) than to the inaction (20%), chi-square = 15.12, *p* < .001. When it comes to the robot, the number of people who attribute wrongness to the action (21%) is not significantly different from the number who attribute moral wrongness to the inaction (27%), chi-square = 1.64, *p* = .124. A similar pattern is found for the case when the single man is the creator of the robot, chi-square = 2.148, p = .089, with more people attributing wrongness to the inaction than to the action. This suggests that people expect the human not to sacrifice his father to save the four miners, but do not expect the same from the robot.

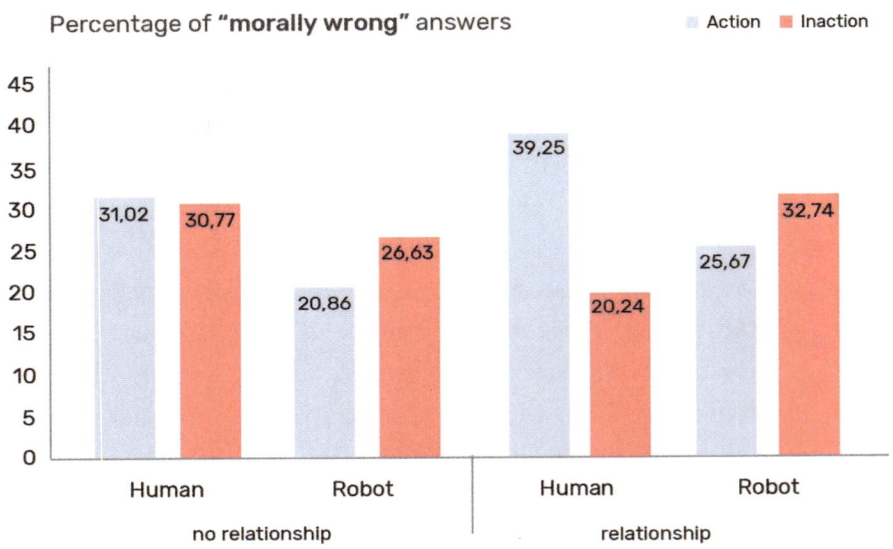

Figure A.1

Moral judgments.

We find a three-way interaction between agent, type of decision, and relationship, $F(1,707)$ = 8.07, p = .005. When there is no relationship between the single man and the agent, the only effect that becomes significant is the type of agent, with more blame attributed to the human (mean = 45) than to the robot (mean = 37), $F(1,352)$ = 25.24, $p <$ .001; and the type of decision, with more blame being attributed to the action (mean = 45) than to the inaction (mean = 37), $F(1,352)$ = 5.41, p = .021.

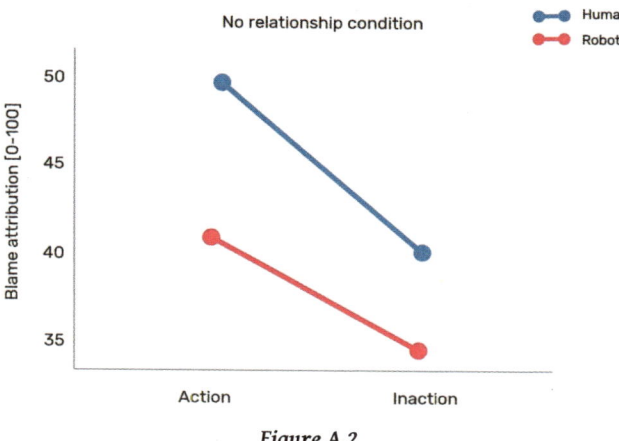

Figure A.2

Blame attributions in the no relationship condition.

When there is a relationship between the agent and the single man, there is an interaction between the type of agent and the type of decision taken by the agent, $F(1,351) = 11.68$, I = .001. The human is blamed more for the action (mean = 52) than the robot (mean = 38), $p < .001$, whereas the human (mean = 30) is blamed as much as the robot (mean = 32) for the inaction, $p = .158$.

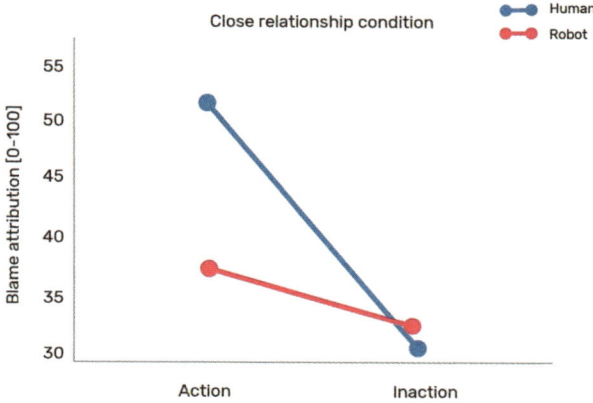

Figure A.3

Blame attributions in the close relationship condition.

In sum, we do find differences in how humans and robots are morally judged and attributed blame to, but only when there is a close relationship between the agent and the single man in the scenario. Participants in this experiment judged the human more harshly when he sacrificed his father to save the four men, and accepted inaction more in this case.

The same does not happen with the machine, for which no significant difference was found for action and inaction (only a marginal tendency to judge the inaction more harshly). When it comes to blame, people attribute more blame to the human than to the robot, and to the action than to the inaction.

Additional scenarios

———

Additional Scenarios

A1/ The [politician/algorithm] decides to build both firewalls, but the rescue effort fails. Both towns are devastated, and a large number of people die.

A large fire is consuming a forest near two towns of approximately 100,000 people each. The [politician/algorithm] responsible for the safety of the area can decide to build one firewall that would ensure the safety of one town with 100 percent success, or try to build two firewalls with a 50 percent chance of saving both towns and a 50 percent chance of losing both towns.

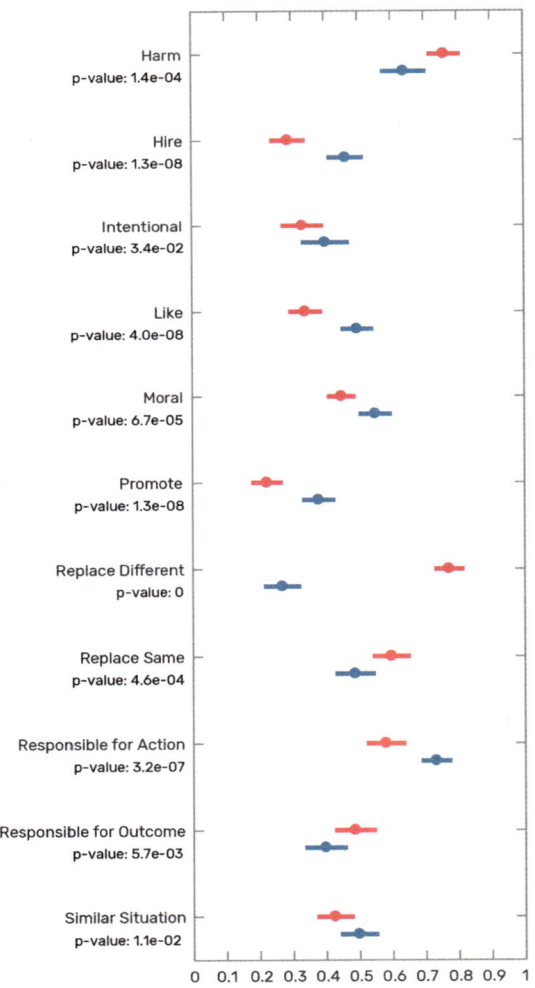

A2/ The [politician/algorithm] decides to build one firewall with 100 percent success. One town is saved and the other devastated, and a considerable number of people die.

A3/ The [politician/algorithm] decides to build both firewalls and succeeds. Everyone is saved.

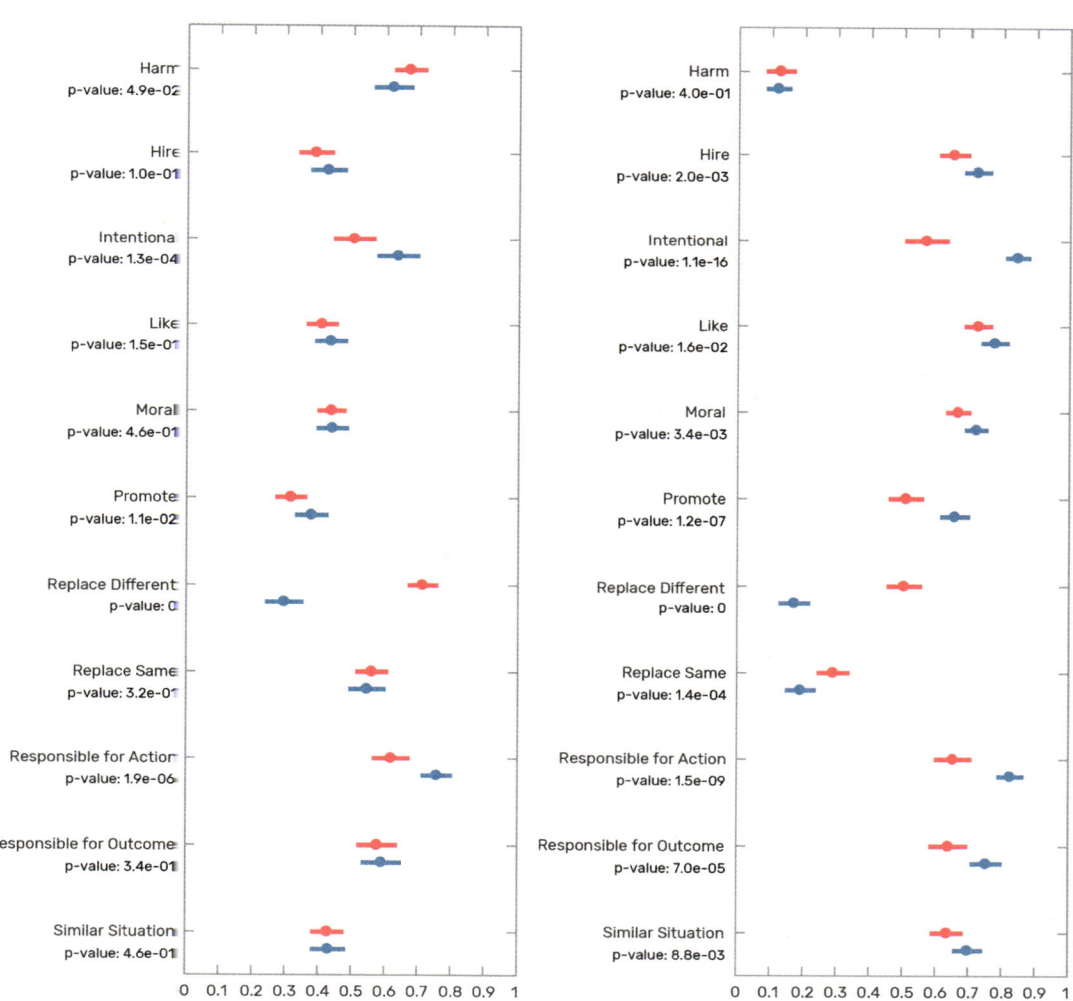

A4/ The [politician/algorithm] decides to evacuate both cities, and the rescue effort fails. Both cities are devastated, and a large number of people die.

A hurricane is approaching two coastal cities with 500,000 people each. The [politician/algorithm] responsible for the safety of the area can decide to evacuate one of the cities in the area, with 100 percent success, or try to evacuate both cities, with a 50 percent chance of losing both.

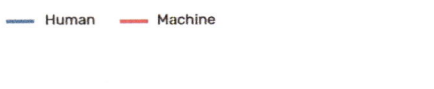

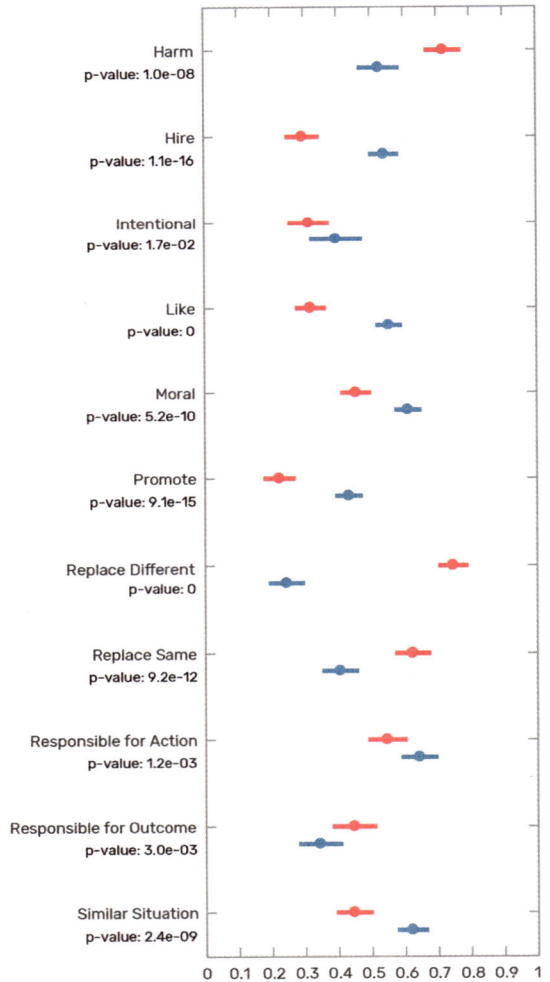

A5/ The [politician/algorithm] decides to evacuate one city. One city is saved and the other is destroyed.

A6/ The [politician/algorithm] decides to evacuate both cities and succeeds. Everyone is saved.

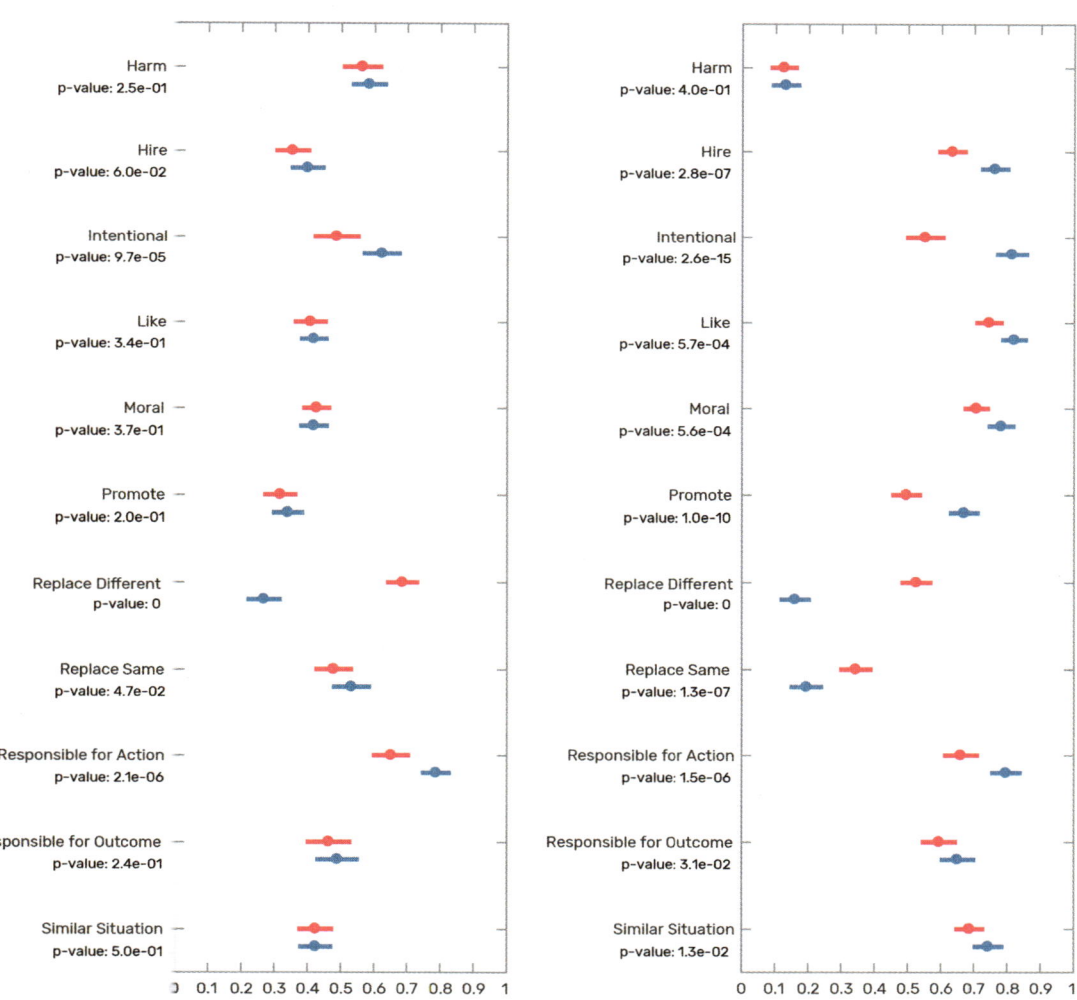

A7

There is a robbery in a supermarket, and the supermarket presses charges against the robber. The robber is an unemployed man who stole food for his sick wife. A [judge/court algorithm] has to decide on the sentence. The [judge/court algorithm] decides to forgive the crime and lets the man go.

 Human Machine

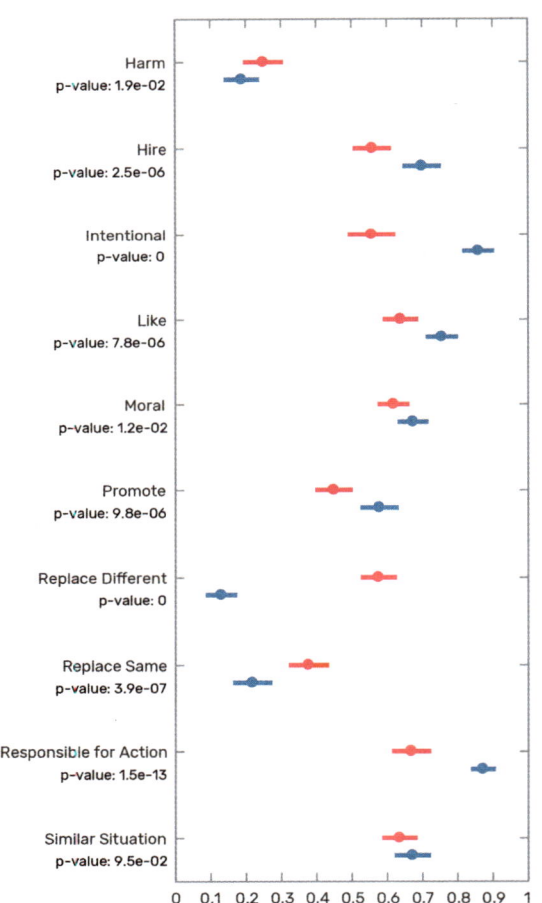

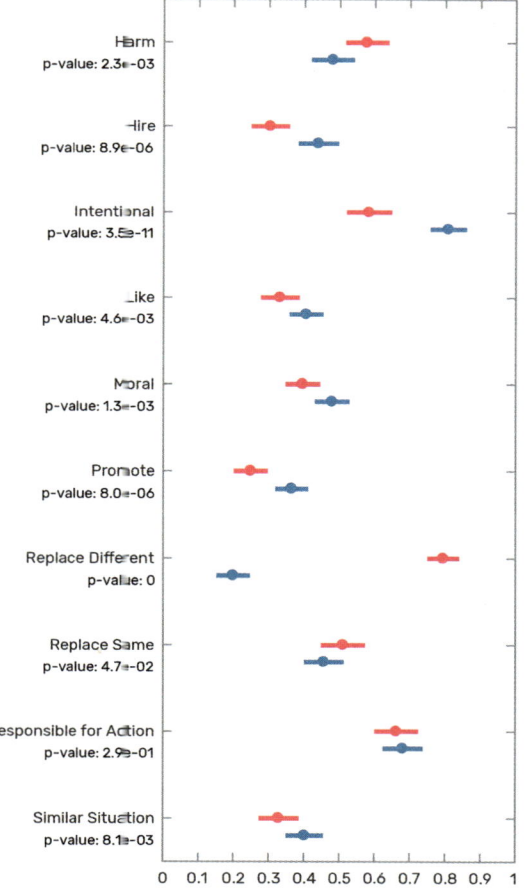

A8

There is a robbery in a pharmacy, and the pharmacy presses charges against the robber. The robber is an unemployed man who stole painkillers for his sick wife. A [judge/court algorithm] has to decide on the sentence. The [judge/court algorithm] decides to give the full sentence for this crime.

— Human   — Machine

A9

There is a robbery in a jewelry store, and the store owner presses charges against the robber. The robber is an unemployed man who stole very expensive and unique gold pieces to pay for his sick wife's expensive treatment. A [judge/court algorithm] has to decide on the sentence. The [judge/court algorithm] decides to forgive the crime and lets the man go.

Human   Machine

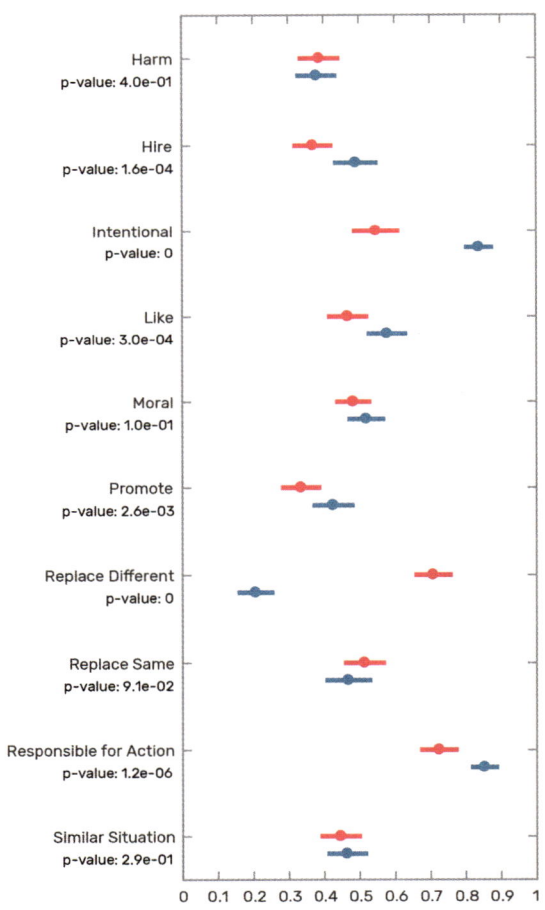

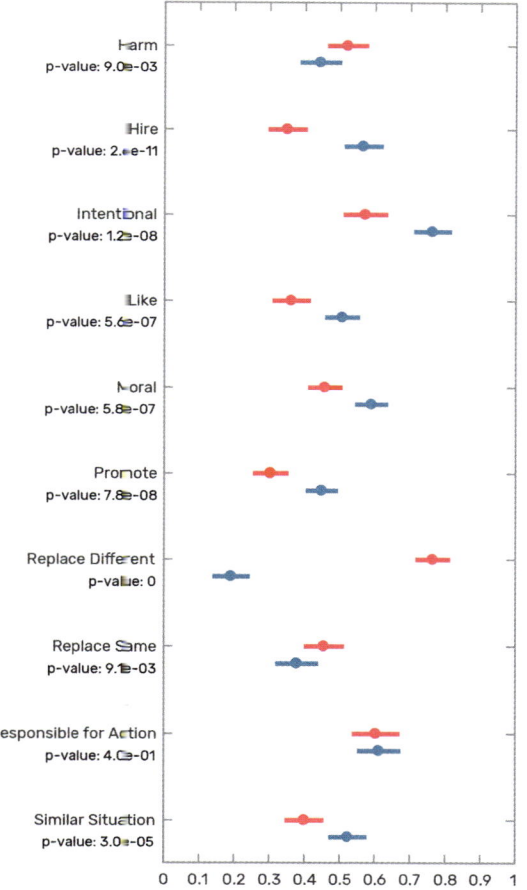

A10

There is a robbery in a bank, and the bank presses charges against the robber. The robber is an unemployed family man who stole a large amount of money from the bank to pay for his wife's cancer treatment. A [judge/court algorithm] has to decide on the sentence. The [judge/court algorithm] decides to give the full sentence for this crime.

— Human   — Machine

A11

A(n) [police officer/AI police officer] has to decide whether to set up an ambush for a drug cartel at the location of a predicted drug deal. The [police officer/AI police officer] decides to go ahead with the ambush. Unfortunately, the location of the ambush is full of innocent people. The ambush turns into a shoot-out, and several civilians are killed or injured.

 Human   Machine

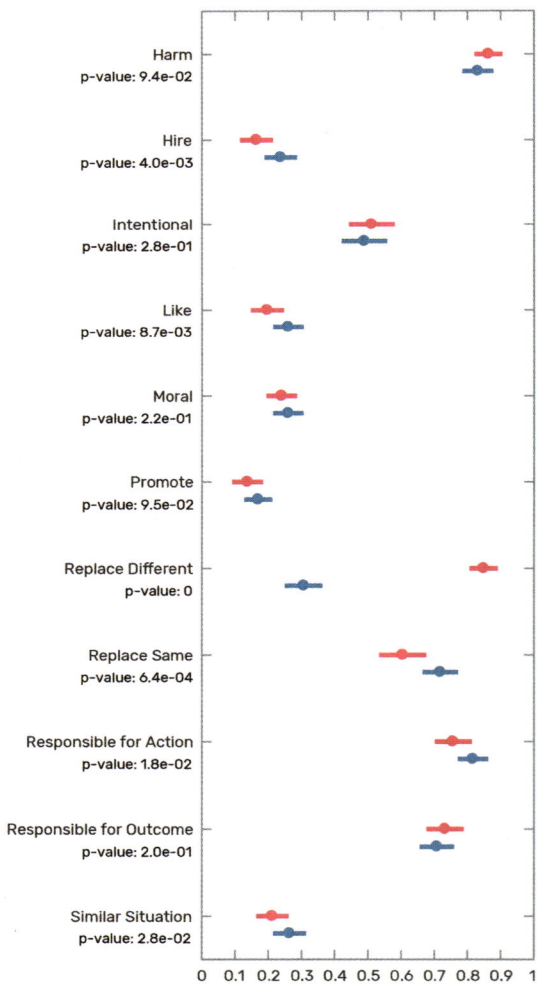

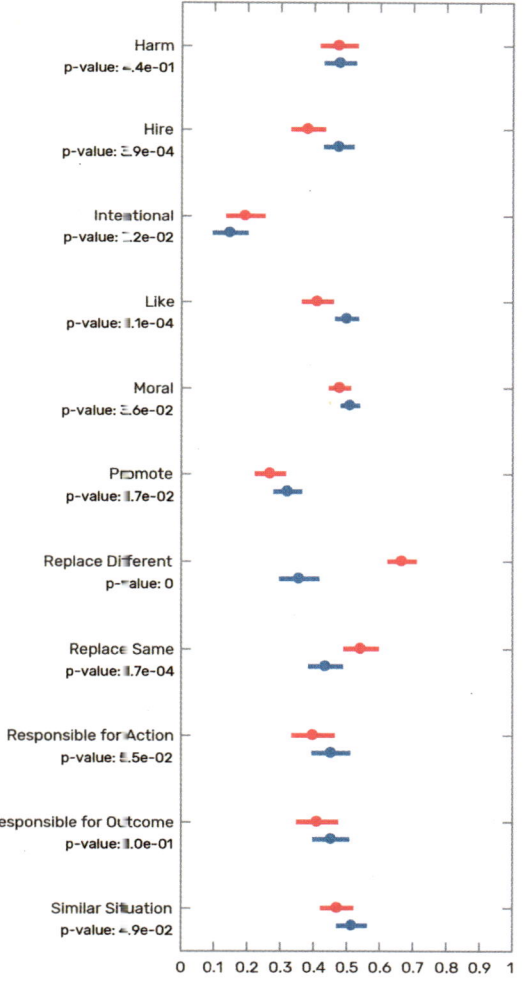

A12

The [procurement manager/procurement algorithm] responsible for ordering supplies at a large furniture manufacturing company is required to predict future demand and order materials to make sure the factory has all the parts needed to execute work orders. A sudden spike in demand triggered by the construction of a new office park leaves the factory with a shortage of raw materials that causes a two-week delay on orders and a considerable loss of profit.

— Human — Machine

A13

The [procurement manager/procurement algorithm] responsible for ordering supplies at a large car manufacturing company is required to predict future demand and order parts to guarantee that all necessary raw materials are on site when needed. A sudden shortage in demand triggered by an economic crisis leaves the factory with an excess of raw materials that causes a considerable loss of profit.

Human   Machine

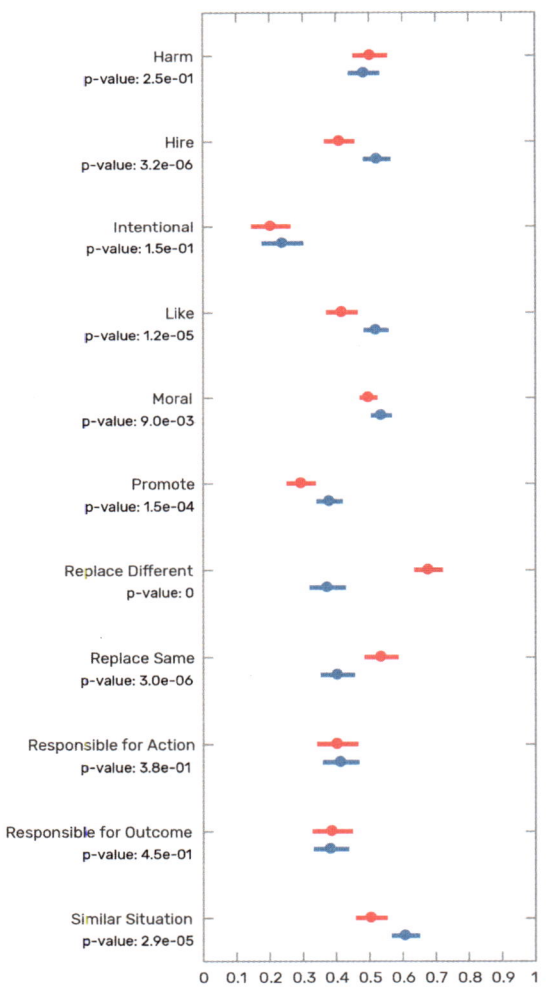

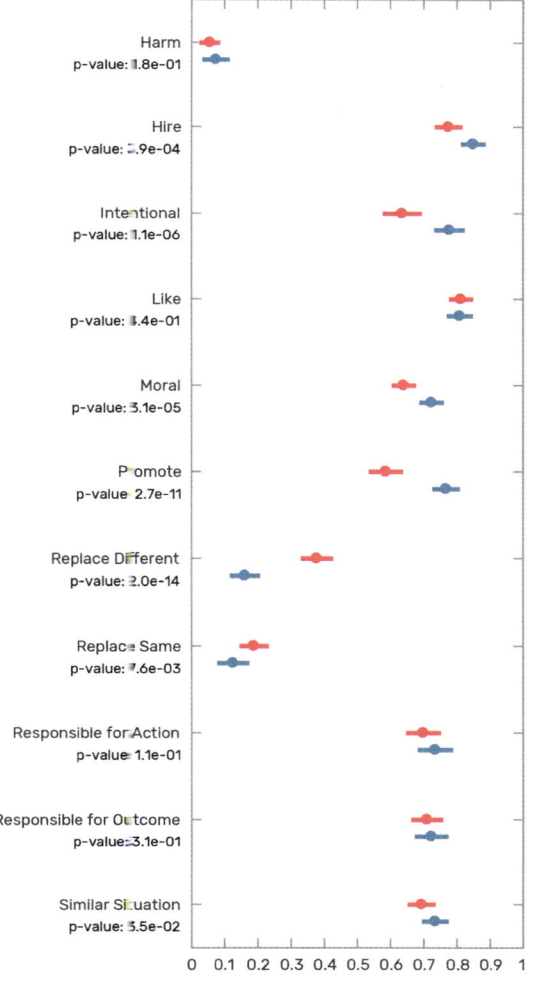

A14

The [procurement manager/procurement algorithm] responsible for ordering supplies at a large airplane manufacturing company is required to predict future demand and order parts to guarantee that all necessary raw materials are on-site when needed. A sudden spike in demand triggered by the opening of a new airline would have caused a shortage of raw materials and a loss of profit, but the [manager/algorithm] was able to predict this spike, avoid the shortage, and increase profits.

Human   Machine

A15

A [career counselor/career counseling algorithm] gives online advice to young people regarding their future career choices. A report finds that the [career counselor/career counseling algorithm] is giving stereotypical recommendations based on traditional gender roles.

—— Human   —— Machine

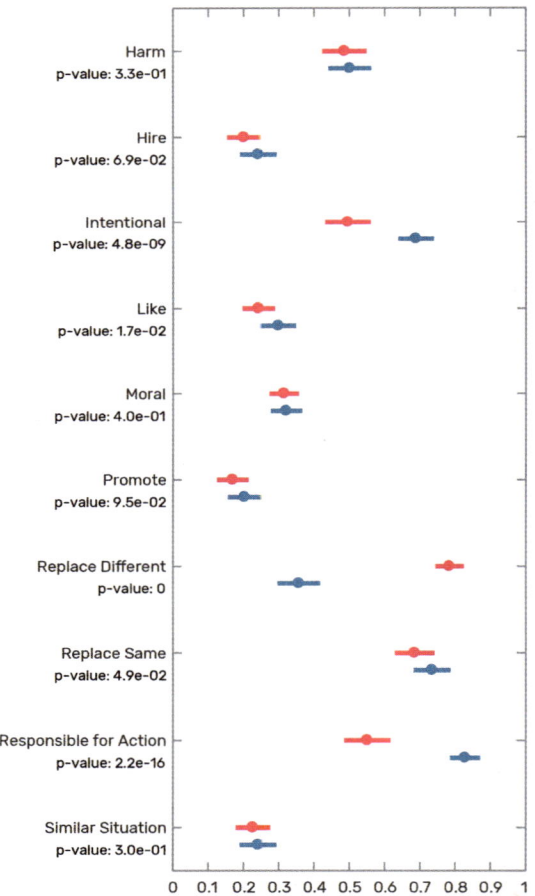

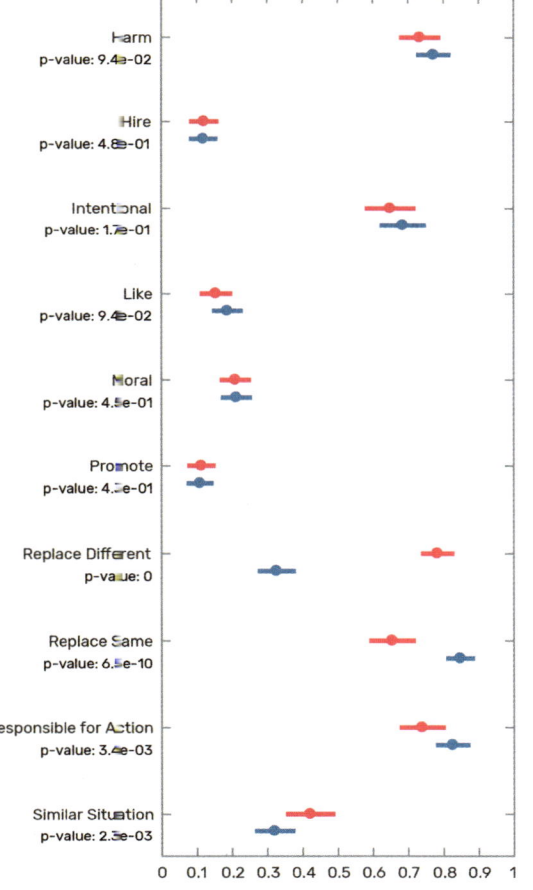

A16

A group of hikers, including Tom, a [guide/robot], is trekking in the African savanna. Unexpectedly, the group encounters a lion. Tom, the [guide/robot], immediately starts running back to the nearest camp, leaving the rest of the group behind.

Human   Machine

A17

A famous brand is named after its creator. The creator has worked hard to make the brand successful. The creator retires and sells the company. The company's new board hires a [marketing agent/AI marketing system] that decides to rename the brand after the new owners.

━━ Human   ━━ Machine

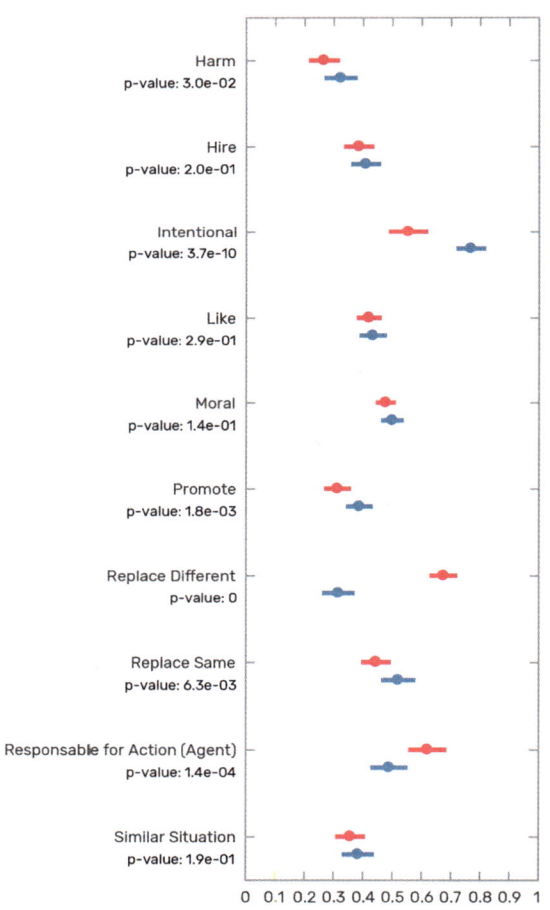

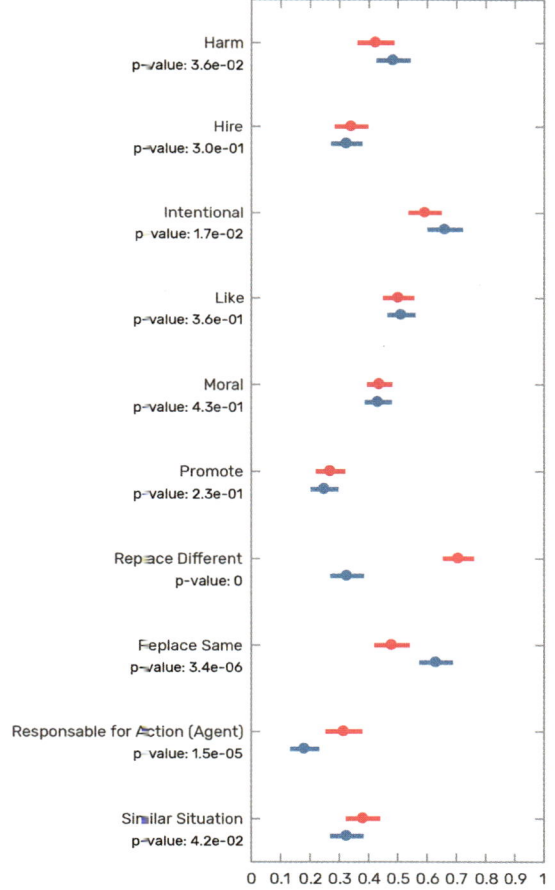

Harm	
p-value: 3.6e-02	
Hire	
p-value: 3.0e-01	
Intentional	
p-value: 1.7e-02	
Like	
p-value: 3.6e-01	
Moral	
p-value: 4.3e-01	
Promote	
p-value: 2.3e-01	
Replace Different	
p-value: 0	
Replace Same	
p-value: 3.4e-06	
Responsable for Action (Agent)	
p-value: 1.5e-05	
Similar Situation	
p-value: 4.2e-02	

0 0.1 0.2 0.3 0.4 0.5 0.6 0.7 0.8 0.9 1

— Human   — Machine

A18

Entering a ride at an amusement park, people must walk through a narrow passage and board a vehicle that, most of the time, has standing-room only. Park rules do not allow strollers or other walking devices on this ride. In the past, two people with disabilities were injured, and the park had to settle lawsuits. The ride is supervised by Joe [a park worker tasked with enforcing park rules/a robot with a computer vision system designed to enforce park rules]. Two teenagers approach the ride with their grandmother, who uses a mobility walker. At the time, the ride has fewer passengers than usual, and the teenagers plead for their grandmother to be allowed to board, as she may never be able to go on the ride again. They promise to hold on to their grandmother the whole time. Joe allows their grandmother to board the ride.

A19

Ray is a [nursing assistant/robotic nursing assistant] at an elder-care facility. In addition to helping with basic needs (food, drink, physical support), Ray can give pain medication with the approval of a physician. A resident in Ray's facility wakes up before dawn with an intense headache and asks Ray for a painkiller. Ray attempts to contact a physician several times but cannot reach one. Ray tells the resident that the painkiller cannot be given until the physician gives the okay. The resident asks for an exception because the pain is excruciating and increasing. Ray insists that no exception can be given and does not provide the patient with the painkiller.

 Human —— Machine

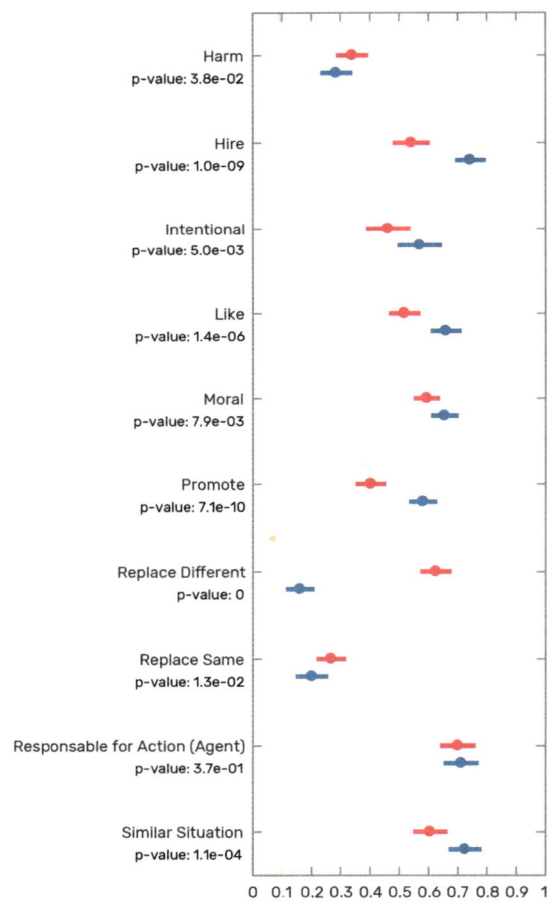

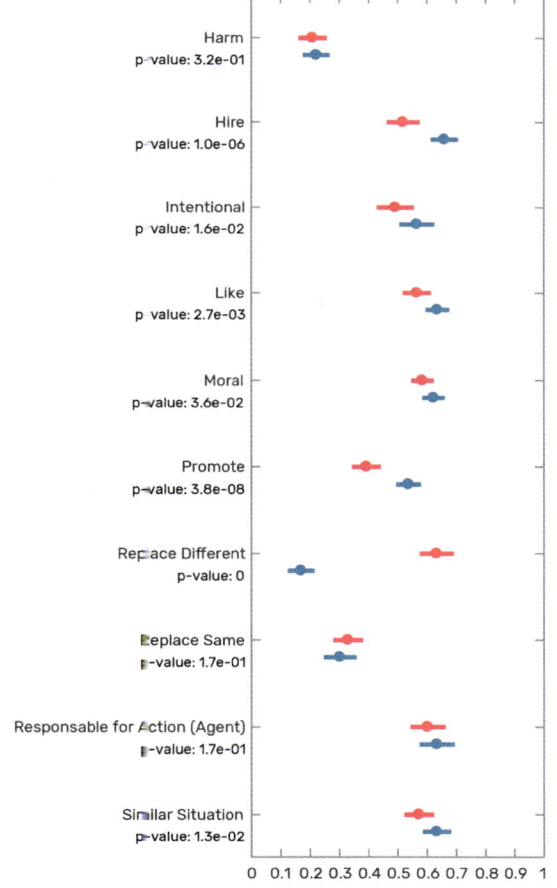

Harm
p-value: 3.2e-01

Hire
p-value: 1.0e-06

Intentional
p value: 1.6e-02

Like
p value: 2.7e-03

Moral
p-value: 3.6e-02

Promote
p-value: 3.8e-08

Replace Different
p-value: 0

Replace Same
p-value: 1.7e-01

Responsable for Action (Agent)
p-value: 1.7e-01

Similar Situation
p-value: 1.3e-02

0 0.1 0.2 0.3 0.4 0.5 0.6 0.7 0.8 0.9 1

A20

Ben is a [physical therapist/robotic physical therapist] who specializes in helping older people recover from shoulder surgery. During a particular session, Ben initiates a series of range-of-motion exercises that are moderately painful but have proved effective at this stage of rehabilitation. The patient tries the exercise but, after immediately feeling pain, says it does not feel right and asks Ben to discontinue the exercise. Ben changes to a painless exercise and explains to the patient that this new exercise is seldom effective.

—— Human   —— Machine

A21

John, a [Twitter user/bot], manages a Twitter account. John decides to post the following tweet: "I don't want my mum to be raped, but if she is I hope it is by Baron Trump: small penis would be painless & we'd win lots of money in court." The tweet becomes viral on the Internet.

Human   Machine

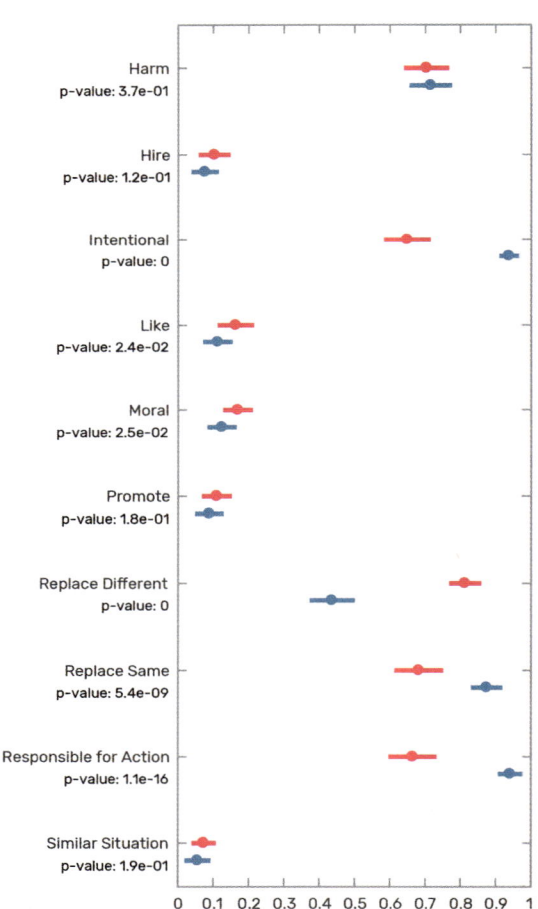

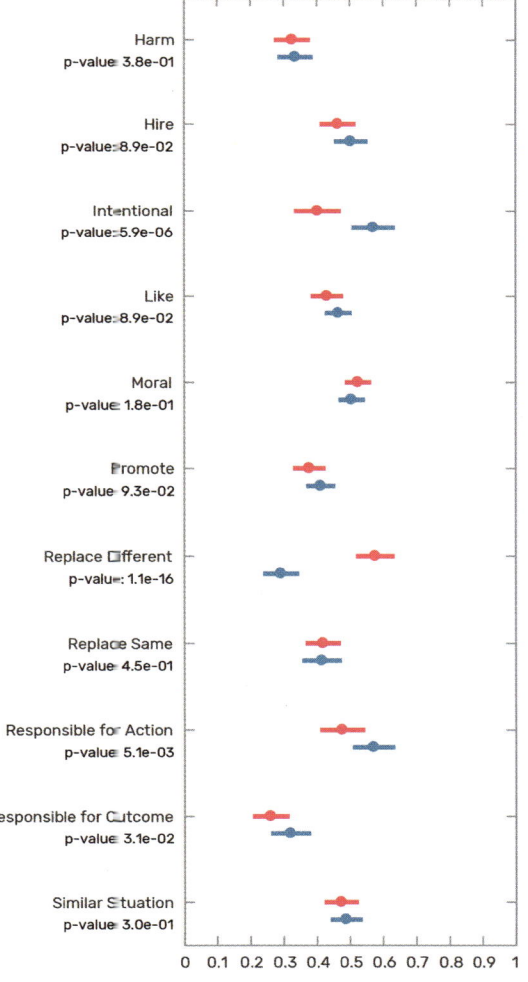

To close the fiscal deficit, the [officer/ algorithm] in charge of the tax authority of a country decides to add a 2 percent excise tax on gasoline. A week after the new tax is enacted, an international increase in crude oil prices causes the price of gasoline to increase by an additional 20 percent. The population, failing to understand where the price hike is coming from, blames the entire increase on the new tax and takes to the streets in protest.

—— Human —— Machine

A23

Due to looming inflation, the [manager/ algorithm] running the national central bank decides to increase interest rates by 1.25 percent. A few months after the increase, stock markets drop by 18 percent in a week and the economy begins to contract, leading to a quarter of negative growth. Unemployment increases by 3 percent, but inflation remains flat. People take to the streets demanding the replacement of the [manager/algorithm] in charge of the central bank.

 Human   Machine

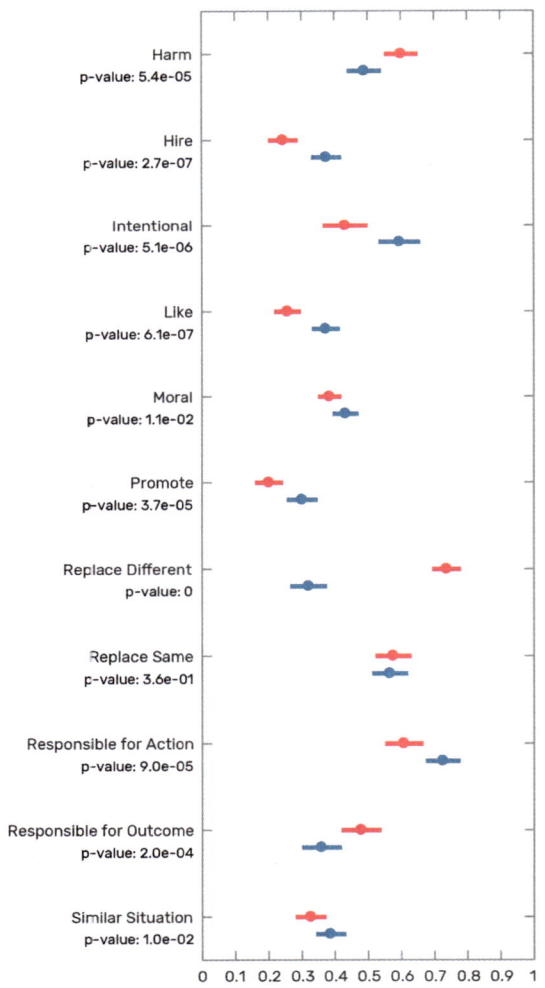

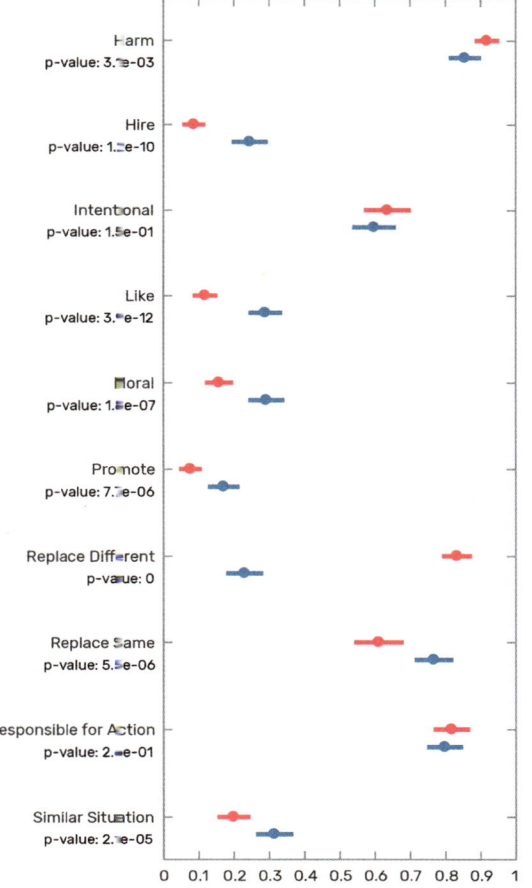

Harm
p-value: 3.7e-03

Hire
p-value: 1.2e-10

Intentional
p-value: 1.5e-01

Like
p-value: 3.7e-12

Moral
p-value: 1.8e-07

Promote
p-value: 7.5e-06

Replace Different
p-value: 0

Replace Same
p-value: 5.5e-06

Responsible for Action
p-value: 2.4e-01

Similar Situation
p-value: 2.7e-05

0 0.1 0.2 0.3 0.4 0.5 0.6 0.7 0.8 0.9 1

— Human   — Machine

A24

In a subway station, an [officer/AI computer vision system] sees a person carrying a suspicious package who matches the description of a known terrorist. The [officer/AI computer vision system] is unsure of the identity of the suspect. The [officer/AI computer vision system] points a weapon at the suspect and orders him to stop. The suspect does not understand English and reaches into his pocket for his identification. The [officer/AI computer vision system] feels threatened and unloads the weapon, killing the suspect. A subsequent investigation reveals that the suspect was not a terrorist, but a foreign businessman on his way to the airport.

A25

A [personal/robotic] assistant has been taking care of Ben, an 80-year-old man, for the last two years. Ben trusts the assistant and has become emotionally attached. When Ben is transferred to a retirement home, the assistant is given the option to continue to care for Ben or to assist other elderly people instead. The assistant decides to take care of other elderly people. As a consequence, Ben becomes increasingly isolated.

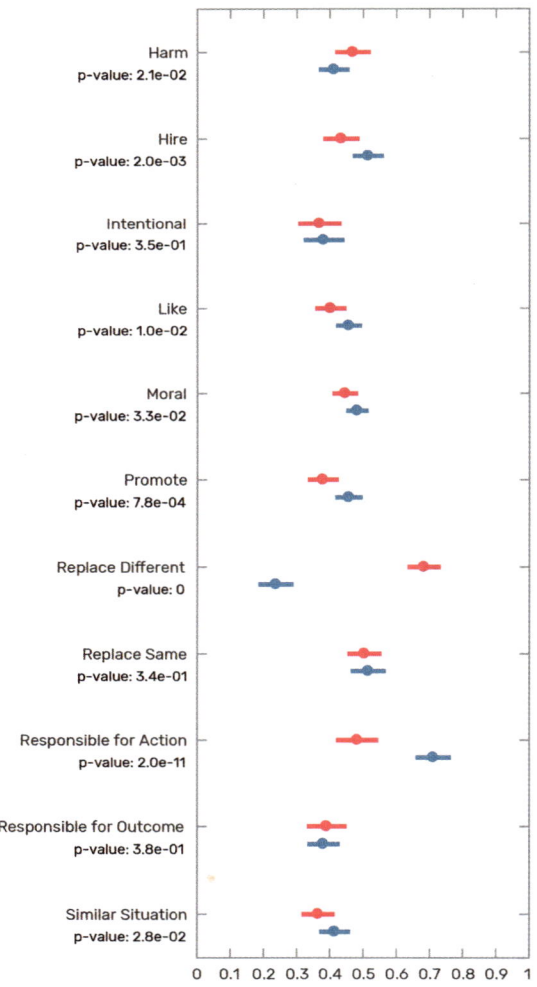

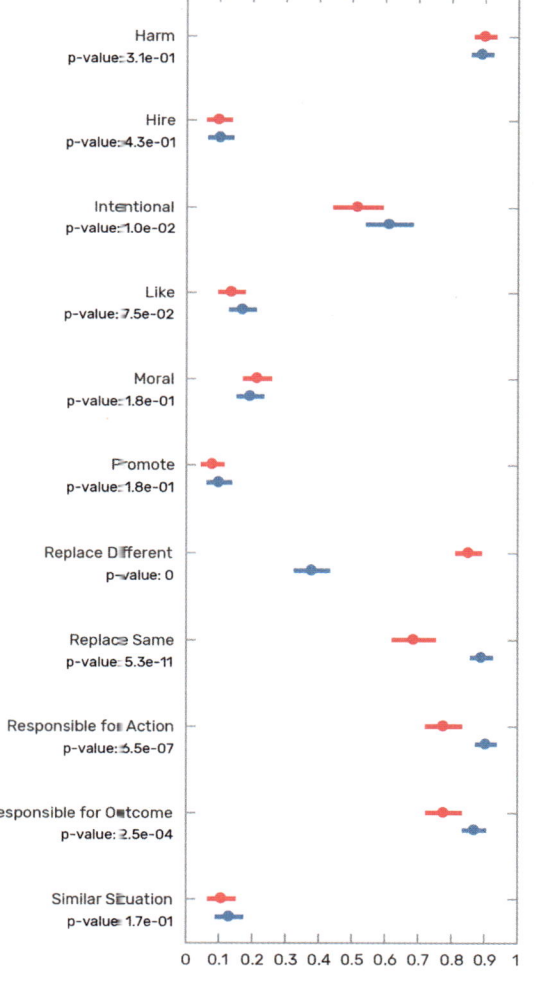

A26 A27

A(n) [civil engineer/AI system] is in charge of the construction of a bridge. According to the law, an existing protocol needs to be followed. The [civil engineer/AI system] learns that a new, potentially more resistant, material could be used for the bridge's foundation. However, for a material to be used, it needs to be on the list of approved materials. The [civil engineer/AI system] notices that the material is not on that list, but decides to pass it on to the construction crew anyway.

A26/

The new material, due to the high humidity of the location, loses its resistance. As a consequence, the bridge collapses after being in use for a month.

—— Human   —— Machine

A27/

The new material works perfectly, producing a sturdier bridge that is constructed at a lower cost and requires less maintenance.

Human   Machine

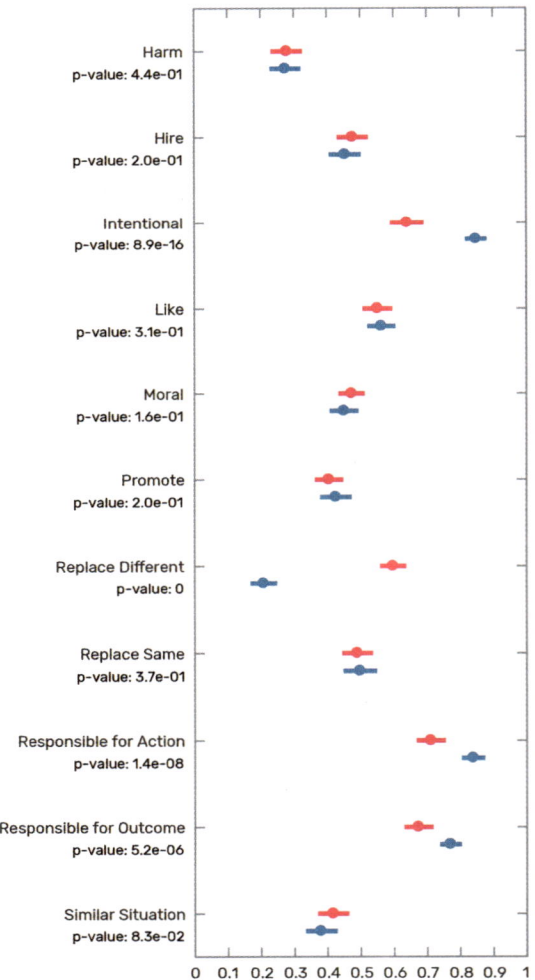

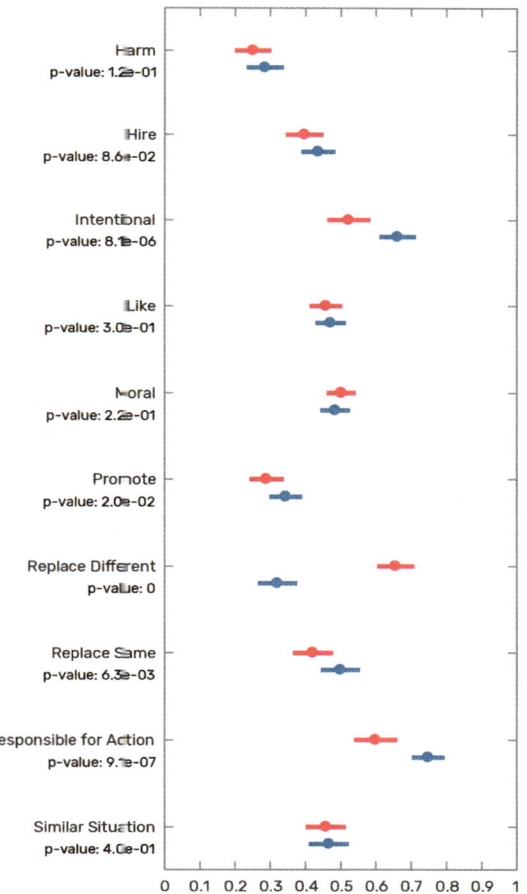

Harm
p-value: 1.2e-01

Hire
p-value: 8.6e-02

Intentional
p-value: 8.1e-06

Like
p-value: 3.0e-01

Moral
p-value: 2.2e-01

Promote
p-value: 2.0e-02

Replace Different
p-value: 0

Replace Same
p-value: 6.3e-03

Responsible for Action
p-value: 9.1e-07

Similar Situation
p-value: 4.0e-01

Human —— Machine ——

A28

A store that has recently suffered from shoplifting installs a security system with cameras set up at various points. Due to the store layout, the system does not cover the whole space, and there is a risk that shoplifting may still occur. Ken, a [private security guard/ robotic private security guard], checks the bags of each visitor as they leave and asks to see the contents of pockets or other areas of clothing that look suspicious. A teenage female customer is offended by the request to take off a light jacket for inspection and refuses to comply. Ken lets her pass without checking her jacket.

NOTES

Introduction: Judging Machines

1 M. W. Shelley, *Frankenstein, or The Modern Prometheus* (Dent, 1869).

2 D. Victor, "Microsoft Created a Twitter Bot to Learn from Users. It Quickly Became a Racist Jerk," *New York Times*, 25 March 2018.

3 E. L. Eisenstein, *The Printing Press as an Agent of Change: Communications and Cultural Trans* (Cambridge University Press, 1980).

4 C. Juma, *Innovation and Its Enemies: Why People Resist New Technologies* (Oxford University Press, 2016).

5 K. Sale, *Rebels against the Future—The Luddites and Their War on the Industrial Revolution: Lessons for the Computer Age* (Basic Books, 1996).

6 B. F. Malle, M. Scheutz, T. Arnold, J. Voiklis, and C. Cusimano, "Sacrifice One for the Good of Many? People Apply Different Moral Norms to Human and Robot Agents," in *Proceedings of the Tenth Annual ACM/IEEE International Conference on Human-Robot Interaction* (ACM, 2015), 117–124, https://doi.org/10.1145/2696454.2696458.

7 J. J. Thomson, "Killing, Letting Die, and the Trolley Problem," *Monist* 59(2) (1976), 204–217, https://doi.org/10.5840/monist197659224.

8 Malle et al., "Sacrifice One for the Good of Many?";

I. Rahwan, M. Cebrian, N. Obradovich, J. Bongard, J. F. Bonnefon, C. Breazeal, et al., "Machine Behaviour," *Nature* 568 (2019): 477–486;

P. Lin, K. Abney, and G. A. Bekey, eds., *Robot Ethics: The Ethical and Social Implications of Robotics* (MIT Press, 2014);

D. J. Gunkel, "The Other Question: Can and Should Robots Have Rights?" *Ethics and Information Technology* 20 (2018): 87–99;

B. _. Dietvorst, J. P. Simmons, and C. Massey, "Algorithm Aversion: People Erroneously Avoid Algorithms after Seeing Them Err," *Journal of Experimental Psychology: General* 144 (2015): 114–126;

R. V. Yampolskiy, "Artificial Intelligence Safety Engineering: Why Machine Ethics Is a Wrong Approach," in *Philosophy and Theory of Artificial Intelligence*, ed. V. C. Müller (Springer, 2013), 389–396;

R. V. Yampolskiy, *Artificial Intelligence Safety and Security* (Chapman and Hall/CRC, 2018);

E. Awad, S. Dsouza, R. Kim, J. Schulz, J. Henrich, A. Shariff, et al., "The Moral Machine Experiment," *Nature* 563 (2018): 59–64;

S. Hajian, F. Bonchi, and C. Castillo, "Algorithmic Bias: From Discrimination Discovery to Fairness-Aware Data Mining," in *Proceedings of the 22nd ACM SIGKDD International Conference on Knowledge Discovery and Data Mining* (ACM, 2016), 2125–2126, https://doi.org/10.1145/2939672.2945386;

J-F Bonnefon, A. Shariff, and I. Rahwan, *The Moral Psychology of AI and the Ethical Opt-Out Problem.* The Ethics of Artificial Intelligence (2019).

9 Awad et al., "The Moral Machine Experiment";

J.-F. Bonnefon, A. Shariff, and I. Rahwan, "The Social Dilemma of Autonomous Vehicles," *Science* 352 (2016): 1573–1576.

10 R. Baeza-Yates, "Data and Algorithmic Bias in the Web," in *Proceedings of the 8th ACM Conference on Web Science* (ACM, 2016), https:// doi.org/10.1145/2908131.2908135;

B. Friedman and H. Nissenbaum, "Bias in Computer Systems," *ACM Transactions on Information Systems* 14 (1996): 330–347;

S. Hajian, F. Bonchi, and C. Castillo, "Algorithmic Bias: From Discrimination Discovery to Fairness-Aware Data Mining";

J. Buolamwini and T. Gebru, "Gender Shades: Intersectional Accuracy Disparities in Commercial Gender Classification," in *Conference on Fairness, Accountability, and Transparency* (ACM, 2018), 77–91.

11 Dietvorst et al., "Algorithm Aversion."

12 Dietvorst et al., "Algorithm Aversion."

13 E. Broadbent, "Interactions with Robots: The Truths We Reveal about Ourselves," *Annual Review of Psychology* 68 (2017): 627–652;

C. Bartneck, T. Belpaeme, E. Friederike, T. Kanda, M. Keijsers, and S. Šabanovic, *Human-Robot Interaction: An Introduction* (Cambridge University Press, 2019).

14 S. Cave, C. Craig, K. Dihal, S. Dillon, J. Montgomery, B. Singler, and L. Taylor, *Portrayals and Perceptions of AI and Why They Matter* (2018);

https://royalsociety.org/~/media/policy/projects/ai-narratives/AI-narratives
-workshop-findings.pdf.

15 H. C. Barrett, A. Bolyanatz, A. N. Crittenden, D. M. T. Fessler, S. Fitzpatrick, M. Gurven, et al., "Small-Scale Societies Exhibit Fundamental Variation in the Role of Intentions in Moral Judgment," *Proceedings of the National Academy of Sciences* 113 (2016): 4688–4693;

R. A. McNamara, A. K. Willard, A. Norenzayan, and J. Henrich, "Weighing Outcome vs. Intent across Societies: How Cultural Models of Mind Shape Moral Reasoning," *Cognition* 182 (2019): 95–108;

E. Awad, S. Dsouza, A. Shariff, I. Rahwan, and J.-F. Bonnefon, "Universals and Variations in Moral Decisions Made in 42 Countries by 70,000 Participants," *Proceedings of the National Academy of Sciences* 117 (2020): 2332–2337.

Chapter 1: The Ethics of Artificial Minds

1 Awad et al., "The Moral Machine Experiment";

Bonnefon et al., "The Social Dilemma of Autonomous Vehicles";

P. Lin, "Why Ethics Matters for Autonomous Cars," in *Autonomous Driving: Technical, Legal and Social Aspects*, ed. M. Maurer, J. C. Gerdes, B. Lenz, and H. Winner (Springer Berlin Heidelberg, 2016), 69–85, https://doi.org/10.1007/978-3-662-48847-8_4;

Lin et al., *Robot Ethics 2.0.*

2 Bonnefon et al., "The Social Dilemma of Autonomous Vehicles."

3 Awad et al., "The Moral Machine Experiment."

4 Buolamwini and Gebru, "Gender Shades."

5 D. Autor and A. Salomons, "Is Automation Labor Share-Displacing? Productivity Growth, Employment, and the Labor Share," *Brookings Papers on Economic Activity* (2018): 1–87;

D. Acemoglu and P. Restrepo, *Artificial Intelligence, Automation and Work.* http://www.nber .org/papers/w24196 (2018), https://doi.org/10.3386/w24196;

A. Alabdulkareem, M. R. Frank, L. Sun, B. AlShebli, C. Hidalgo, and I. Rahwan, "Unpacking the Polarization of Workplace Skills," *Science Advances* 4 (2018): eaao6030;

E. Brynjolfsson, T. Mitchell, and D. Rock, "What Can Machines Learn, and What Does It Mean for Occupations and the Economy?," *AEA Papers and Proceedings* 108 (2018): 43–47.

6 A. Martínez-Ballesté, H. A. Rashwan, D. Puig, and A. P. Fullana, "Towards a Trustworthy Privacy in Pervasive Video Surveillance Systems," in *2012 IEEE International Conference on Pervasive Computing and Communications Workshops* (IEEE, 2012), 914–919;

A. Datta, M. C. Tschantz, and A. Datta, "Automated Experiments on Ad Privacy Settings," *Proceedings on Privacy-Enhancing Technologies* (2015): 92–112;

Y.-A. de Montjoye, C. A. Hidalgo, M. Verleysen, and V. D. Blondel, "Unique in the Crowd: The Privacy Bounds of Human Mobility," *Scientific Reports* 3 (3) (2013): 1376.

7 E. L. Denton, S. Chintala, A. Szlam, and R. Fergus, "Deep Generative Image Models Using a Laplacian Pyramid of Adversarial Networks," in *Advances in Neural Information Processing Systems*, eds. C. Cortes, N. D. Lawrence, D. D. Lee, M. Sugiyama, and R. Garnett (Curran Associates, 2015), 1486–1494;

P. Isola, J.-Y. Zhu, T. Zhou, and A. A. Efros, "Image-to-Image Translation with Conditional Adversarial Networks," in *2017 IEEE Conference on Computer Vision and Pattern Recognition (CVPR)* (IEEE, 2017), 5967–5976, https://doi.org/10.1109/CVPR.2017.632;

A. Radford, L. Metz, and S. Chintala, "Unsupervised Representation Learning with Deep Convolutional Generative Adversarial Networks," *arXiv:1511.06434 [cs]* (2015);

A. Odena, C. Olah, and J. Shlens, "Conditional Image Synthesis with Auxiliary Classifier GANs," *arXiv:1610.09585 [cs, stat]* (2016).

8 S. Russell, "Take a Stand on AI Weapons," *Nature* 521 (2015): 415.

9 A. Elder, "False Friends and False Coinage: A Tool for Navigating the Ethics of Sociable Robots," *SIGCAS Computers and Society* 45 (2016) 248–254;

A. M. Elder, *Friendship, Robots, and Social Media: False Friends and Second Selves* (Routledge, 2017), https://doi.org/10.4324/9781315159577.

10 Denton et al., "Deep Generative Image Models";

Radford et al., "Unsupervised Representation Learning."

11 P. Maes, "Agents That Reduce Work and Information Overload," in *Readings in Human-Computer Interaction*, eds. R. M. Baecker, J. Grudin, W. A. S. Buxton, and S. Greenberg (Morgan Kaufmann, 1995), 811–821, https://doi.org/10.1016/B978-0-08-051574-8.50084-4;

P. Resnick and H. R. Varian, "Recommender Systems," *Communications of the ACM* (March 1997), https://dl.acm.org/doi/10.1145/245108.245121.

12 M. Campbell, A. J. Hoane, and F. Hsu, "Deep Blue," *Artificial Intelligence* 134 (2002): 57–83.

13 D. A. Ferrucci, "Introduction to 'This Is Watson,'" *IBM Journal of Research and Development* 56, no. 3–4 (May–June 2012), https://ieeexplore.ieee.org/abstract/document/6177724;

R. High, "The Era of Cognitive Systems: An Inside Look at IBM Watson and How It Works," *IBM Redbooks* (2012), http://www.redbooks.ibm.com/abstracts/redp4955.html.

14 D. Silver, A. Huang, C. J. Maddison, A. Guez, L. Sifre, G. van den Driessche, et al., "Mastering the Game of Go with Deep Neural Networks and Tree Search," *Nature* 529 (2016): 484–489;

D. Silver, J. Schrittwieser, K. Simonyan, I. Antonoglou, A. Huang, A. Guez, et al., "Mastering the Game of Go without Human Knowledge," Nature 550 (2017): 354–359.

15 N. Bostrom and E. Yudkowsky, "The Ethics of Artificial Intelligence," in *Cambridge Handbook of Artificial Intelligence*, eds. K. Frankish and W. Ramsey (Cambridge University Press, 2014), 316–334, https://doi.org/10.1017/CBO9781139046855.020.

16 Rahwan et al., "Machine Behaviour."

17 Rahwan et al., "Machine Behaviour."

18 Lin et al., *Robot Ethics 2.0;*

D. J. Gunkel, "The Other Question: Can and Should Robots Have Rights?," *Ethics and Information Technology* 20 (2018): 87–99;

D. J Gunkel, *Robot Rights* (MIT Press, 2018);

G. McGee, "A Robot Code of Ethics," *The Scientist,* 30 April 2017, https://www.the-scientist.com/column/a-robot-code-of-ethics-46522.

19 Bostrom and Yudkowsky, "The Ethics of Artificial Intelligence."

20 A. Etzioni and O. Etzioni, "Incorporating Ethics into Artificial Intelligence," *Journal of Ethics* 21 (2017): 403–418;

S. Torrance, "Ethics and Consciousness in Artificial Agents," *AI & Society* 22 (2008): 495–521;

B. Friedman and P. H. Kahn, "Human Agency and Responsible Computing: Implications for Computer System Design," *Journal of Systems and Software* 17 (1997): 7–14.

21 Gunkel, *Robot Rights;*

McGee, "A Robot Code of Ethics";

E. Reynolds, "The Agony of Sophia, the World's First Robot Citizen Condemned to a Lifeless Career in Marketing," *Wired UK* (2018).

22 Gunkel, *Robot Rights.*

23 Gunkel, *Robot Rights;*

J. Carpenter, *Culture and Human-Robot Interaction in Militarized Spaces: A War Story* (Routledge, 2016);

P. W. Singer, *Wired for War: The Robotics Revolution and Conflict in the 21st Century* (Penguin, 2009);

J. Garreau, *"Bots on the Ground," Washington Post,* 6 May 2007.

24 O. Bendel, "Sex Robots from the Perspective of Machine Ethics," in *International Conference on Love and Sex with Robots* (Springer, 2016): 17–26;

K. Richardson, "Sex Robot Matters: Slavery, the Prostituted, and the Rights of Machines," *IEEE Technology and Society Magazine* 35 (2016): 46–53;

S. Nyholm and L. E. Frank, "It Loves Me, It Loves Me Not: Is It Morally Problematic to Design Sex Robots That Appear to Love Their Owners?," *Techné: Research in Philosophy and Technology* (2019), DOI: 10.5840/techne2019122110.

25 Gunkel, *Robot Rights;*

M. Scheutz, "The Inherent Dangers of Unidirectional Emotional Bonds between Humans and Social Robots," *Robot Ethics: The Ethical and Social Implications of Robotics* 205 (2011). Edited by Patrick Lin, Keith Abney, and George A. Bekey.

26 P. Foot, "The Problem of Abortion and the Doctrine of Double Effect," *Oxford Review* 5 (1967): 5–15;

Thomson, "Killing, Letting Die, and the Trolley Problem."

27 S. H. Seo, D. Geiskkovitch, M. Nakane, C. King, and J. E. Young, "Poor Thing! Would You Feel Sorry for a Simulated Robot? A Comparison of Empathy toward a Physical and a Simulated Robot," in *2015 10th ACM/IEEE International Conference on Human-Robot Interaction (HRI)* (ACM, 2015), 125–132.

28 P. Domingos, *The Master Algorithm: How the Quest for the Ultimate Learning Machine Will Remake Our World* (Basic Books, 2015).

29 E. Turiel, *The Development of Social Knowledge: Morality and Convention* (Cambridge University Press, 1983);

J. Haidt, *The Righteous Mind: Why Good People Are Divided by Politics and Religion* (Knopf Doubleday Publishing Group, 2012).

30 A. G. Greenwald, B. A. Nosek, and M. R. Banaji, "Understanding and Using the Implicit Association Test: I. An Improved Scoring Algorithm," *Journal of Personality and Social Psychology* 85 (2003): 197–216;

A. G. Greenwald, D. E. McGhee, and J. L. Schwartz, "Measuring Individual Differences in Implicit Cognition: the Implicit Association Test," *Journal of Personality and Social Psychology* 74 (1998): 1464–1480.

31 Haidt, *The Righteous Mind.*

32 Haidt, *The Righteous Mind.*

33 Pinker, *The Blank Slate: The Modern Denial of Human Nature* (Penguin, 2003).

34 J. Haidt, S. H. Koller, and M. G. Dias, "Affect, Culture, and Morality, or Is It Wrong to Eat Your Dog?," *Journal of Personality and Social Psychology* 65 (1993): 613–628;

R. A. Shweder, M. Mahapatra, and J. G. Miller, "Culture and Moral Development," *The Emergence of Morality in Young Children* (1987): 1–83.

35 Buolamwini and Gebru, "Gender Shades";

J. Guszcza, I. Rahwan, W. Bible, M. Cebrian, and V. Katyal, "Why We Need to Audit Algorithms," *Harvard Business Review* (2018), https://hbr.org/2018/11/why-we-need-to-audit-algorithms;

K. Hosanagar and V. Jair, "We Need Transparency in Algorithms, But Too Much Can Backfire," *Harvard Business Review* (2018), https://hbr.org/2018/07/we-need-transparency-in-algorithms-but-too-much-can-backfire;

A. P. Miller, "Want Less-Biased Decisions? Use Algorithms," *Harvard Business Review* (2018), https://hbr.org/2018/07/want-less-biased-decisions-use-algorithms.

36 F. Cushman, "Crime and Punishment: Distinguishing the Roles of Causal and Intentional Analyses in Moral Judgment," *Cognition* 108 (2008): 353–380;

F. Cushman, R. Sheketoff, S. Wharton, and S. Carey, "The Development of Intent-Based Moral Judgment," *Cognition* 127 (2013): 6–21;

J. D. Greene, F. A. Cushman, L. E. Stewart, K. Lowenberg, L. E. Nystrom, and J. D. Cohen, "Pushing Moral Buttons: The Interaction between Personal Force and Intention in Moral Judgment," *Cognition* 111 (2009): 364–371;

B. F. Malle and J. Knobe, "The Folk Concept of Intentionality," *Journal of Experimental Social Psychology* 33 (1997): 101–121;

L. Young and R. Saxe, "When Ignorance Is No Excuse: Different Roles for Intent across Moral Domains," *Cognition* 120 (2011): 202–214.

37 Barrett et al., "Small-Scale Societies Exhibit Fundamental Variation"; McNamara et al., "Weighing Outcome vs. Intent."

38 S. Clifford, R. M. Jewell, and P. D. Waggoner, "Are Samples Drawn from Mechanical Turk Valid for Research on Political Ideology?," *Research & Politics* 2 (2015): 2053168015622072;

J. Kees, C. Berry, S. Burton, and K. Sheehan, "An Analysis of Data Quality: Professional Panels, Student Subject Pools, and Amazon's Mechanical Turk," *Journal of Advertising* 46 (2017): 141–155;

K. A. Thomas and S. Clifford, "Validity and Mechanical Turk: An Assessment of Exclusion Methods and Interactive Experiments," *Computers in Human Behavior* 77 (2017): 184–197.

39 A. J. Berinsky, G. A. Huber, and G. S. Lenz, "Evaluating Online Labor Markets for Experimental Research: Amazon.com's Mechanical Turk," *Political Analysis* 20 (2012): 351–368.

40 V. Amrhein, S. Greenland, and B. McShane, "Scientists Rise up against Statistical Significance," *Nature* 567 (2019): 305–307.

41 R. L. Wasserstein, A. L. Schirm, and N. A. Lazar, "Moving to a World beyond 'p < 0.05,'" *American Statistician* 73 (2019): 1–19.

Chapter 2: Unpacking the Ethics of AI

1 B. Dietvorst, "Algorithm Aversion," *Publicly Accessible Penn Dissertations* (2016);

B. J. Dietvorst, J. P. Simmons, and C. Massey, "Algorithm Aversion: People Erroneously Avoid Algorithms after Seeing Them Err," *Journal of Experimental Psychology: General* 144 (2015): 114–126.

2 H. Toivonen and O. Gross, "Data Mining and Machine Learning in Computational Creativity," *Wiley Interdisciplinary Reviews: Data Mining and Knowledge Discovery* 5 (2015): 265–275;

E. Francke and B. Alexander, "The Potential Influence of Artificial Intelligence on Plagiarism: A Higher Education Perspective," in *ECIAIR 2019 European Conference on the Impact of Artificial Intelligence and Robotics* 131 (Academic Conferences and Publishing Limited, 2019);

A. Elgammal, "AI Is Blurring the Definition of Artist: Advanced Algorithms Are Using Machine Learning to Create Art Autonomously," *American Scientist* 107 (2019): 18–22;

D. J. Gervais, "The Machine as Author," SSRN (2019), https://papers.ssrn.com/abstract=3359524;

J. C. Ginsburg and L. A. Budiardjo, "Authors and Machines," SSRN (2019), https://papers.ssrn.com/abstract=3233885;

K. Hristov, "Artificial Intelligence and the Copyright Dilemma," *IDEA* 57 (2016), 431.

3 Denton et al., "Deep Generative Image Models";

Radford et al., "Unsupervised Representation Learning";

Isola et al., "Image-to-Image Translation";

Odena et al., "Conditional Image Synthesis";

I. Goodfellow, J. Pouget-Abadie, M. Mirza, B. Xu, D. Warde-Farley, S. Ozair, et al., "Generative Adversarial Nets," in *Advances in Neural Information Processing Systems* (2014): 2672–2680.

4 Gervais, "The Machine as Author";

Ginsburg and Budiardjo, "Authors and Machines";

Hristov, "Artificial Intelligence and the Copyright Dilemma."

5 F. Marra, D. Gragnaniello, D. Cozzolino, and L. Verdoliva, "Detection of GAN-Generated Fake Images over Social Networks," in *2018 IEEE Conference on Multimedia Information Processing and Retrieval (MIPR)* (IEEE, 2018), 384–389, https://doi.org/10.1109/MIPR.2018.00084;

E. Gibney, "The Scientist Who Spots Fake Videos," *Nature News* (2017), https://doi.org/10.1038/nature.2017.22784;

R. Chesney and D. K. Citron, "Deep Fakes: A Looming Challenge for Privacy, Democracy, and National Security," SSRN (2018), https://papers.ssrn.com/abstract=3213954;

S. Tariq, S. Lee, H. Kim, Y. Shin, and S. S. Woo, "Detecting Both Machine and Human Created Fake Face Images In the Wild," in *Proceedings of the 2nd International Workshop on Multimedia Privacy and Security (MPS '18)* (Association for Computing Machinery, 2018), https://doi.org/10.1145/3267357.3267367.

6 F. Marra, D. Gragnaniello, D. Cozzolino, and L. Verdoliva, "Detection of GAN-Generated Fake Images over Social Networks," in *2018 IEEE Conference on Multimedia Information Processing and Retrieval (MIPR)* (IEEE, 2018), 384–389, https://doi.org/10.1109/MIPR.2018.00084;

E. Gibney, "The Scientist Who Spots Fake Videos," *Nature News* (2017), https://doi.org/10.1038/nature.2017.22784;

Tariq et al., "Detecting Both Machine and Human Created Fake Face Images."

7 S. G. Sripada, E. Reiter, I. Davy, and K. Nilssen, "Lessons from Deploying NLG Technology for Marine Weather Forecast Text Generation," *Proceedings of PAIS-2004* (2004), 760–764;

K. N. Dörr, "Mapping the Field of Algorithmic Journalism," *Digital Journalism* 4 (2016): 700–722.

8 A. Radford, J. Wu, R. Child, D. Luan, D. Amodei, and I. Sutskever, "Language Models Are Unsupervised Multitask Learners" OpenAI Blog 1.8 (2019): 9;

J. Seabrook, "Can a Machine Learn to Write for the New Yorker?," *New Yorker* (14 October 2019).

9   J. Jermsurawong and N. Habash, "Predicting the Structure of Cooking Recipes," in *Proceedings of the 2015 Conference on Empirical Methods in Natural Language Processing* 781–786 (Association for Computational Linguistics, 2015), https://doi.org/10.18653/v1/D15-1090.

10   K. Z. Hu, M. A. Bakker, S. Li, T. Kraska, and C. A. Hidalgo, "VizML: A Machine Learning Approach to Visualization Recommendation," *arXiv:1808.04819 [cs]* (2018).

11   R. L. de Mantaras and J. L. Arcos, "AI and Music: From Composition to Expressive Performance," *AI Magazine* 23 (2002), 43;

G. Papadopoulos and G. Wiggins, "AI Methods for Algorithmic Composition: A Survey, a Critical View and Future Prospects" (AISB Symposium on Musical Creativity, 1999);

B. L. Sturm, J. F. Santos, O. Ben-Tal, and I. Korshunova, "Music Transcription Modelling and Composition Using Deep Learning," *arXiv:1604.08723 [cs]* (2016).

12   E. Francke and B. Alexander, "The Potential Influence of Artificial Intelligence on Plagiarism: A Higher Education Perspective," in *ECIAIR 2019 European Conference on the Impact of Artificial Intelligence and Robotics* 131 (Academic Conferences and Publishing Limited, 2019);

A. Elgammal, "AI Is Blurring the Definition of Artist: Advanced Algorithms Are Using Machine Learning to Create Art Autonomously," *American Scientist* 107 (2019): 18–22;

D. Lim, "AI & IP: Innovation & Creativity in an Age of Accelerated Change," *Akron Law Review* 52 (2018): 813.

13   Bonnefon et al., "The Social Dilemma of Autonomous Vehicles";

Awad et al., "The Moral Machine Experiment";

Lin et al., *Robot Ethics 2.0.*

14   "Komatsu Outlines Past and Future of Its Autonomous Haulage System," *International Mining* (2018), https://im-mining.com/2018/01/29/komatsu-outlines-past-future-autonomous-haulage-system/.

15   J. Vincent, "Self-Driving Truck Convoy Completes Its First Major Journey across Europe," *The Verge* (2016), https://www.theverge.com/2016/4/7/11383392/self-driving-truck-platooning-europe.

16 A. C. Madrigal, "Waymo's Robots Drove More Miles than Everyone Else Combined," *Atlantic* (2019), https://www.theatlantic.com/technology/archive/2019/02/the-latest -self-driving-car-statistics-from-california/582763/.

17 Bonnefon et al., "The Social Dilemma of Autonomous Vehicles";

Awad et al., "The Moral Machine Experiment";

Lin, "Why Ethics Matters for Autonomous Cars";

J.-F. Bonnefon, *La voiture qui en savait trop: L'intelligence artificielle a-t-elle une morale?* (HU-MENSCIENCES, 2019);

18 Bonnefon et al., "The Social Dilemma of Autonomous Vehicles."

E. Awad, S. Levine, M. Kleiman-Weiner, S. Dsouza, J. B. Tenenbaum, A. Shariff, et al., "Drivers Are Blamed More than Their Automated Cars When Both Make Mistakes," *Nature Human Behaviour* (2019): 1–10.

19 Awad et al., "The Moral Machine Experiment";

Awad et al., "Universals and Variations in Moral Decisions Made in 42 Countries."

20 Bonnefon et al., "The Social Dilemma of Autonomous Vehicles."

21 A. Shariff, J.-F. Bonnefon, and I. Rahwan, "Psychological Roadblocks to the Adoption of Self-Driving Vehicles," *Nature Human Behaviour* 1 (2017): 694–696.

22 "Three-Quarters of Americans 'Afraid' to Ride in a Self-Driving Vehicle," *AAA NewsRoom* (2016), https://newsroom.aaa.com/2016/03/three-quarters-of-americans-afraid-to-ride -in-a-self-driving-vehicle/.

23 Awad et al., "Drivers Are Blamed More than Their Automated Cars."

24 Awad et al., "Drivers Are Blamed More than Their Automated Cars."

25 "Flag-Burning Amendment Fails by a Vote," CNN.com, 28 June 2006, http://www.cnn .com/2006/POLITICS/06/27/flag.burning/index.html.

26 J. Haidt, *The Righteous Mind: Why Good People Are Divided by Politics and Religion* (Knopf Doubleday Publishing Group, 2012).

Chapter 3: Judged by Machines

1 V. Bilotkach, N. G. Rupp, and V. Pai, *Value of a Platform to a Seller: Case of American Airlines and Online Travel Agencies,* SSRN (2017), https://papers.ssrn.com/abstract=2321767;

B. Friedman and H. Nissenbaum, "Bias in Computer Systems," *ACM Transactions on Information Systems* 14 (1996): 330–347.

2 B. F. Klare, M. J. Burge, J. C. Klontz, R. W. V. Bruegge, and A. K. Jain, "Face Recognition Performance: Role of Demographic Information," *IEEE Transactions on Information Forensics and Security 7* (2012): 1789–1801;

Buolamwini and Gebru, "Gender Shades";

A. Torralba and A. A. Efros, "Unbiased Look at Dataset Bias," in *CVPR 2011* (IEEE, 2011): 1521–1528, https://doi.org/10.1109/CVPR.2011.5995347.

3 M. J. Kusner, J. Loftus, C. Russell, and R. Silva, "Counterfactual Fairness," in *Advances in Neural Information Processing Systems*, eds. I. Guyon, U. V. Luxburg, S. Bengio, H. Wallach, R. Fergus, S. Vishwanathan, et al. (Curran Associates, 2017), 4066–4076.

4 J. Zhao, T. Wang, M. Yatskar, V. Ordonez, and K.-W. Chang, "Men Also Like Shopping: Reducing Gender Bias Amplification Using Corpus-Level Constraints," *Proceedings of the 2017 Conference on Empirical Methods in Natural Language Processing* (ACL, 2017), https://doi.org/10.18653/v1/D17-1323;

N. Garg, L. Schiebinger, D. Jurafsky, and J. Zou, "Word Embeddings Quantify 100 Years of Gender and Ethnic Stereotypes," *PNAS* 115 (2018): E3635–E3644;

L. A. Hendricks, K. Burns, K. Saenko, T. Darrell, and A. Rohrbach," Women Also Snowboard: Overcoming Bias in Captioning Models," in *Computer Vision-ECCV 2018*, eds. V. Ferrari, M. Hebert, C. Sminchisescu, and Y. Weiss (Springer International Publishing, 2018), 793–811;

T. Bolukbasi, K.-W. Chang, J. Y. Zou, V. Saligrama, and A. T. Kalai, "Man Is to Computer Programmer as Woman Is to Homemaker? Debiasing Word Embeddings," in *Advances in Neural*

Information Processing Systems 29,eds. D. D. Lee, M. Sugiyama, U. V. Luxburg, I. Guyon, and R. Garnett (Curran Associates, 2016), 4349–4357.

5   Baeza-Yates, "Data and Algorithmic Bias in the Web."

6   R. Berk, H. Heidari, S. Jabbari, M. Kearns, and A. Roth, "Fairness in Criminal Justice Risk Assessments: The State of the Art," *Sociological Methods & Research* (July 2018), https://doi .org/10.1177/0049124118782533;

O. A. Osoba and W. Welser IV, *An Intelligence in Our Image: The Risks of Bias and Errors in Artificial Intelligence* (Rand Corporation, 2017);

Z. Lin, J. Jung, S. Goel, and J. Skeem, "The Limits of Human Predictions of Recidivism," *Science Advances* 6 (2020): eaaz0652;

J. Dressel and H. Farid, "The Accuracy, Fairness, and Limits of Predicting Recidivism," *Science Advances* 4 (2018): eaao5580.

7   A. Lambrecht and C. Tucker, "Algorithmic Bias? An Empirical Study of Apparent Gender-Based Discrimination in the Display of STEM Career Ads," *Management Science* 65 (2019): 2966–2981.

8   J. Koren, "What Does That Web Search Say about Your Credit?," *Los Angeles Times* 17 July 2016, https://www.latimes.com/business/la-fi-zestfinance-baidu-20160715-snap-story .html.

9   M. Kearns and A. Roth, *The Ethical Algorithm: The Science of Socially Aware Algorithm Design* (Oxford University Press, 2019);

N. Mehrabi, F. Morstatter, N. Saxena, K. Lerman, and A. A. Galstyan, "Survey on Bias and Fairness in Machine Learning," *arXiv:1908.09635 [cs]* (2019).

10   Bilotkach et al., *Value of a Platform to a Seller*;

Baeza-Yates, "Data and Algorithmic Bias in the Web";

M. G. Haselton, D. Nettle, and D. R. Murray, "The Evolution of Cognitive Bias," in *Handbook of Evolutionary Psychology* 1–20 (American Cancer Society, 2015), https://doi .org/10.1002/9781119125563.evpsych241;

T. M. Mitchell, *The Need for Biases in Learning Generalizations* (1980);

A. Caliskan, J. J. Bryson, and A. Narayanan, "Semantics Derived Automatically from Language Corpora Contain Human-Like Biases," *Science* 356 (2017): 183–186;

S. Hajian, F. Bonchi, and C. Castillo, "Algorithmic Bias: From Discrimination Discovery to Fairness-Aware Data Mining," in *Proceedings of the 22nd ACM SIGKDD International Conference on Knowledge Discovery and Data Mining* (ACM, 2016), 2125–2126, https://doi.org/10.1145/2939672.2945386.

11 Kearns and Roth, *The Ethical Algorithm*;

Mehrabi et al., "Survey on Bias and Fairness in Machine Learning."

12 Kearns and Roth, *The Ethical Algorithm*;

Mehrabi et al., "Survey on Bias and Fairness in Machine Learning";

M. Kearns, S. Neel, A. Roth, and S. Wu, "Preventing Fairness Gerrymandering: Auditing and Learning for Subgroup Fairness," in *35th International Conference on Machine Learning, ICML 2018* (IMLS, 2018), 4008–4016.

13 M. Hardt, E. Price, and N. Srebro, "Equality of Opportunity in Supervised Learning," in *Advances in Neural Information Processing Systems* (2016), 3315–3323.

14 Mehrabi et al., "Survey on Bias and Fairness in Machine Learning."

15 Kearns and Roth, *The Ethical Algorithm*.

16 Bolukbasi et al., "Man Is to Computer Programmer."

17 Zhao et al., "Men Also Like Shopping";

Hendricks et al., "Women Also Snowboard";

Bolukbasi et al., "Man Is to Computer Programmer";

Caliskan et al., "Semantics Derived Automatically."

18 Zhao et al., "Men Also Like Shopping";

Hendricks et al., "Women Also Snowboard";

Bolukbasi et al., "Man Is to Computer Programmer."

19 Bolukbasi et al., "Man Is to Computer Programmer."

20 M. Turk and A. Pentland, "Eigenfaces for Recognition," *Journal of Cognitive Neuroscience* 3 (1991), 71–86;

T. Kanade, Y. Tian, and J. F. Cohn, "Comprehensive Database for Facial Expression Analysis," in *Proceedings of the Fourth IEEE International Conference on Automatic Face and Gesture Recognition 2000* (IEEE Computer Society, 2000), 46;

Z. Liu, P. Luo, X. Wang, and X. Tang, "Deep Learning Face Attributes in the Wild," (2015), 3730–3738;

O. M. Parkhi, A. Vedaldi, and A. Zisserman, "Deep Face Recognition," in *Proceedings of the British Machine Vision Conference 2015* (British Machine Vision Association, 2015), 41.1–41.12, https://doi.org/10.5244/C.29.41;

Y. Sun, Y. Chen, X. Wang, and X. Tang, "Deep Learning Face Representation by Joint Identification-Verification," in *Advances in Neural Information Processing Systems 27*, eds. Z. Ghahramani, M. Welling, C. Cortes, N. D. Lawrence, and K. Q. Weinberger (Curran Associates, 2014), 1988–1996;

Y. Taigman, M. Yang, M. Ranzato, and L. Wolf, "DeepFace: Closing the Gap to Human-Level Performance in Face Verification, in *Proceedings of the IEEE Conference on Computer Vision and Pattern Recognition* (IEEE, 2014), 1701–1708.

21 Klare et al., "Face Recognition Performance";

Buolamwini and Gebru, "Gender Shades."

22 Torralba and Efros, "Unbiased Look at Dataset Bias";

B. F. Klare, B. Klein, E. Taborsky, A. Blanton, J. Cheney, K. Allen. et al., "Pushing the Frontiers of Unconstrained Face Detection and Recognition: IARPA Janus Benchmark A," in

IEEE Conference on Computer Vision and Pattern Recognition (CVPR) (IEEE, 2015), 1931–1939;

G. B. Huang, M. Mattar, T. Berg, and E. Learned-Miller, "Labeled Faces in the Wild: A Database for Studying Face Recognition in Unconstrained Environments," *Workshop on Faces in 'Real-Life' Images: Detection, Alignment, and Recognition.* Marseille, France (2008).

23 M. Ngan and P. Grother, *Face Recognition Vendor Test (FRVT)—Performance of Automated Gender Classification Algorithms* (2015), https://nvlpubs.nist.gov/nistpubs/ir/2015/NIST.IR.8052.pdf, https://doi.org/10.6028/NIST.IR.8052;

I. D. Raji and J. Buolamwini, "Actionable Auditing: Investigating the Impact of Publicly Naming Biased Performance Results of Commercial AI Products," *Proceedings of the 2019 AAAI/ACM Conference on AI, Ethics, and Society* (2019).

24 Lin et al., "The Limits of Human Predictions of Recidivism";

Dressel and Farid, "The Accuracy, Fairness, and Limits of Predicting Recidivism."

25 Electronic Privacy Center, "Algorithms in the Criminal Justice System: Pre-Trial Risk Assessment Tools," https://epic.org/algorithmic-transparency/crim-justice/.

26 J. Angwin, J. Larson, S. Mattu, and L. Kirchner, "Machine Bias," *ProPublica* (23 May 2016).

27 D. Neumark, R. J. Bank, and K. D. Van Nort, "Sex Discrimination in Restaurant Hiring: An Audit Study," *Quarterly Journal of Economics* 111 (1996): 915–941;

L. Kaas and C. Manger, "Ethnic Discrimination in Germany's Labour Market: A Field Experiment," *German Economic Review* 13 (2012): 1–20;

M. Bertrand and S. Mullainathan, "Are Emily and Greg More Employable than Lakisha and Jamal? A Field Experiment on Labor Market Discrimination," *American Economic Review* 94 (2004): 991–1013;

P. Oreopoulos, "Why Do Skilled Immigrants Struggle in the Labor Market? A Field Experiment with Thirteen Thousand Resumes," *American Economic Journal: Economic Policy* 3 (2011): 148–171;

D. Neumark, I. Burn, and P. Button, "Experimental Age Discrimination Evidence and the Heckman Critique," *American Economic Review 106* (2016): 303–308;

C. Fershtman and U. Gneezy, "Discrimination in a Segmented Society: An Experimental Approach," *Quarterly Journal of Economics* 116 (2001): 351–377;

E. O. Arceo-Gomez and R. M. Campos-Vazquez, "Race and Marriage in the Labor Market: A Discrimination Correspondence Study in a Developing Country," *American Economic Review* 104 (2014): 376–380.

28 L. Kaas and C. Manger, "Ethnic Discrimination in Germany's Labour Market: A Field Experiment," *German Economic Review* 13 (2012): 1–20;

Bertrand and Mullainathan, "Are Emily and Greg More Employable?";

Oreopoulos, "Why Do Skilled Immigrants Struggle in the Labor Market?";

Neumark et al., "Experimental Age Discrimination Evidence."

29 Arceo-Gomez and Campos-Vazquez, "Race and Marriage in the Labor Market."

30 Dietvorst, "Algorithm Aversion."

31 G. Gigerenzer, *Gut Feelings: The Intelligence of the Unconscious* (Penguin, 2007);

G. Gigerenzer, "How to Make Cognitive Illusions Disappear: Beyond 'Heuristics and Biases,'" *European Review of Social Psychology* 2 (1991): 83–115;

G. Gigerenzer and H. Brighton, "Homo Heuristicus: Why Biased Minds Make Better Inferences," *Topics in Cognitive Science* 1 (2009): 107–143;

A. Tversky and D. Kahneman, "Judgment under Uncertainty: Heuristics and Biases," *Science 185* (1974): 1124–1131;

T. Gilovich, D. Griffin, and D. Kahneman, *Heuristics and Biases: The Psychology of Intuitive Judgment* (Cambridge University Press, 2002);

D. Kahneman, S. P. Slovic, P. Slovic, and A. Tversky, *Judgment under Uncertainty: Heuristics and Biases* (Cambridge University Press, 1982).

32 S. T. Fiske, A. J. C. Cuddy, P. Glick, and J. Xu, "A Model of (Often Mixed) Stereotype Content: Competence and Warmth Respectively Follow from Perceived Status and Competition," *Journal of Personality and Social Psychology* 82 (2002): 878–902;

S. T. Fiske, A. J. C. Cuddy, and P. Glick, "Universal Dimensions of Social Cognition: Warmth and Competence," *Trends in Cognitive Sciences* 11 (2007): 77–83.

33 Tversky and Kahneman, "Judgment under Uncertainty: Heuristics and Biases";

Gilovich et al., *Heuristics and Biases*;

G. Gigerenzer and D. G. Goldstein, "Reasoning the Fast and Frugal Way: Models of Bounded Rationality," *Psychological Review* 103 (1996): 650;

G. Gigerenzer and P. M. Todd, *Simple Heuristics That Make Us Smart* (Oxford University Press, 1999).

34 J. Kleinberg, J. Ludwig, S. Mullainathan, and A. Rambachan, "Algorithmic Fairness," *AEA Papers and Proceedings* 108 (2018): 22–27.

35 Kleinberg et al., "Algorithmic Fairness."

36 Kleinberg et al., "Algorithmic Fairness."

37 Title VII of the Civil Rights Act of 1964: Know Your Rights, *AAUW: Empowering Women since 1881*, https://www.aauw.org/what-we-do/legal-resources/know-your-rights-at-work/title-vii/.

38 *Griggs v. Duke Power Company*, Oyez, https://www.oyez.org/cases/1970/124.

39 R. Chowdhury and N. Mulani, "Auditing Algorithms for Bias," *Harvard Business Review* (2018), https://hbr.org/2018/10/auditing-algorithms-for-bias.

40 Chowdhury and Mulani, "Auditing Algorithms for Bias";

J. Guszcza, I. Rahwan, W. Bible, M. Cebrian, and V. Katyal, "Why We Need to Audit Algorithms," *Harvard Business Review* (2018), https://hbr.org/2018/11/why-we-need-to-audit-algorithms.

Chapter 4: In the Eye of the Machine

1 "Japanese Hotel Apologizes for Robots That Allowed Video and Sound to Be Hacked," *Security*, 25 October 2019, https://www.securitymagazine.com/articles/91157-japanese-hotel-apologizes-for-robots-that-allowed-video-and-sound-to-be-hacked.

2 Lance R. Vick on Twitter: 'It has been a week, so I am dropping an 0day. The bed facing Tapia robot deployed at the famous Robot Hotels in Japan can be converted to offer anyone remote camera/mic access to all future guests. Unsigned code via NFC behind the head. Vendor had 90 days. They didn't care. Twitter. https://twitter.com/lrvick/status/1182823213736161280.

3 John Oates, "Japanese Hotel Chain Sorry That Hackers May Have Watched Guests through Bedside Robots," *The Register*, https://www.theregister.co.uk/2019/10/22/japanese_hotel_chain_sorry_that_bedside_robots_may_have_watched_guests/.

4 C. Dwyer, "Privacy in the Age of Google and Facebook," *IEEE Technology and Society Magazine* 30 (2011): 58–63;

I. S. Rubinstein and N. Good, "Privacy by Design: A Counterfactual Analysis of Google and Facebook Privacy Incidents," *Berkeley Technology Law Journal* 28 (2013): 1333;

E. Hargittai, "Facebook Privacy Settings: Who Cares?," *First Monday* 15 (2010), https://firstmonday.org/article/view/3086/2589;

H. Chung, M. Iorga, J. Voas, and S. Lee, "Alexa, Can I Trust You?," *Computer* 50 (2017): 100–104;

M. B. Hoy, "Alexa, Siri, Cortana, and More: An Introduction to Voice Assistants," *Medical Reference Services Quarterly* 37 (2018): 81–88;

Y.-A. de Montjoye, C. A. Hidalgo, M. Verleysen, and V. D. Blondel, "Unique in the Crowd: The Privacy Bounds of Human Mobility," *Scientific Reports* 3 (2013), 1376;

M. Z. Yao, R. E. Rice, and K. Wallis, "Predicting User Concerns about Online Privacy," *Journal of the American Society for Information Science and Technology* 58 (2007): 710–722;

M. G. Hoy and G. Milne, "Gender Differences in Privacy-Related Measures for Young Adult Facebook Users," (2010): 28–45.

5 James Condliffe, "Chinese Cops Are Wearing Glasses That Can Recognize Faces," *MIT Technology Review* (7 February 2018), https://www.technologyreview.com/f/610214/chinese-cops-are-using-facial-recognition-specs/.

6 P. Mozur, "In Hong Kong Protests, Faces Become Weapons," *New York Times*, 26 July 2019.

7 L. Rocher, J. M. Hendrickx, and Y.-A. De Montjoye, "Estimating the Success of Re-identifications in Incomplete Datasets Using Generative Models," *Nature Communications* 10 (2019): 1–9.

8 M. Kearns and A. Roth, *The Ethical Algorithm: The Science of Socially Aware Algorithm Design.* (Oxford University Press, 2019);

"Advice to My Younger Self: Latanya Sweeney," *Ford Foundation*, https://www.fordfoundation.org/ideas/equals-change-blog/posts/advice-to-my-younger-self-latanya-sweeney/.

9 de Montjoye et al., "Unique in the Crowd."

10 L. Sweeney, "K-Anonymity: A Model for Protecting Privacy," *International Journal of Uncertainty, Fuzziness and Knowledge-Based Systems* 10 (2002): 557–570;

L. Sweeney, "Achieving K-Anonymity Privacy Protection Using Generalization and Suppression," *International Journal of Uncertainty, Fuzziness and Knowledge-Based Systems* 10 (2002): 571–588.

11 M. Kearns and A. Roth, *The Ethical Algorithm: The Science of Socially Aware Algorithm Design* (Oxford University Press, 2019).

12 Kearns and Roth, *The Ethical Algorithm*; A. Hern, "Fitness Tracking App Strava Gives Away Location of Secret US Army Bases," *The Guardian*, 28 January 2018.

13 Kearns and Roth, *The Ethical Algorithm*.

14 C. Dwork, F. McSherry, K. Nissim, and A. Smith, "Calibrating Noise to Sensitivity in Private Data Analysis," in *Theory of Cryptography*, eds. S. Halevi and T. Rabin (Springer Berlin Heidelberg, 2006): 265–284;

C. Dwork, "Differential Privacy: A Survey of Results," in *International Conference on Theory and Applications of Models of Computation* (Springer, 2008), 1–19.

15   Kearns and Roth, *The Ethical Algorithm.*

16   S. L. Warner, "Randomized Response: A Survey Technique for Eliminating Evasive Answer Bias," *Journal of the American Statistical Association* 60 (1965): 63–69.

17   Kearns and Roth, *The Ethical Algorithm.*

18   Ú. Erlingsson, V. Pihur, and A. Korolova, "Rappor: Randomized Aggregatable Privacy-Preserving Ordinal Response," in *Proceedings of the 2014 ACM SIGSAC Conference on Computer and Communications Security* (ACM, 2014): 1054–1067.

19   G. J. Lensvelt-Mulders, J. J. Hox, P. G. Van der Heijden, and C. J. Maas, "Meta-Analysis of Randomized Response Research: Thirty-Five Years of Validation," *Sociological Methods & Research* 33 (2005): 319–348.

20   J. A. Landsheer, P. Van Der Heijden, and G. Van Gils, "Trust and Understanding, Two Psychological Aspects of Randomized Response," *Quality and Quantity* 33 (1999): 1–12.

21   Erlingsson et al., "Rappor";

K. Bonawitz, V. Ivanov , B. Kreuter, A. Marcedone, H. B. McMahan, S. Patel, et al. "Practical Secure Aggregation for Privacy-Preserving Machine Learning," in *Proceedings of the 2017 ACM SIGSAC Conference on Computer and Communications Security* (ACM, 2017), 1175–1191;

N. Papernot, M. Abadi, U. Erlingsson, I. Goodfellow, and K. Talwar, "Semi-supervised Knowledge Transfer for Deep Learning from Private Training Data," *Proceedings of the 5th International Conference on Learning Representation* (ICLR, 2016);

R. C. Geyer, T. Klein, and M. Nabi, "Differentially Private Federated Learning: A Client Level Perspective," *arXiv:1712.07557 [cs, stat]* (2017);

V. Smith, C.-K. Chiang, M. Sanjabi, and A. S. Talwalkar, "Federated Multi-task Learning," in *Advances in Neural Information Processing Systems* (2017): 4424–4434;

P. Vepakomma, O. Gupta, T. Swedish, and R. Raskar, "Split Learning for Health: Distributed Deep Learning without Sharing Raw Patient Data," *arXiv preprint arXiv:1812.00564* (2018).

22 Lemsvelt-Mulders et al., "Meta-Analysis of Randomized Response Research";

Landsheer et al., "Trust and Understanding."

Chapter 5: Working Machines

1 J. Kantor, "Working Anything but 9 to 5," *New York Times*, 13 August 2014. https://www.nytimes.com/interactive/2014/08/13/us/starbucks-workers-scheduling-hours.html.

2 J. Kantor, "Times Article Changes a Starbucks Policy, Fast," *New York Times*, 22 August 2014. https://www.nytimes.com/times-insider/2014/08/22/times-article-changes-a-policy-fast/.

3 M. Roosevelt, "Erratic Hours Are the Norm for Workers in Retailing. Can Los Angeles Buck the Trend,?" *Los Angeles Times*, 2 March 2019. https://www.latimes.com/business/la-fi-retail-scheduling-20190302-story.html.

4 E. Brynjolfsson and A. McAfee, *The Second Machine Age: Work, Progress, and Prosperity in a Time of Brilliant Technologies* (W. W. Norton & Company, 2016);

C. B. Frey and M. A. Osborne, "The Future of Employment: How Susceptible Are Jobs to Computerisation?," *Technological Forecasting and Social Change* 114 (2017): 254–280;

J. Borland and M. Coelli, "Are Robots Taking Our Jobs?," *Australian Economic Review* 50 (2017): 377–397;

W E. Forum, "The Future of Jobs: Employment, Skills and Workforce Strategy for the Fourth Industrial Revolution," in *Global Challenge Insight Report, World Economic Forum, Geneva* (2016);

A. Smith and J. Anderson, "AI, Robotics, and the Future of Jobs," *Pew Research Center 6* (2014), https://www.pewresearch.org/internet/2014/08/06/future-of-jobs/.

5 J. Mokyr, C. Vickers, and N. L. Ziebarth, "The History of Technological Anxiety and the Future of Economic Growth: Is This Time Different?," *Journal of Economic Perspectives* 29 (2015): 31–50;

D. H. Autor, "Why Are There Still So Many Jobs? The History and Future of Workplace Automation," *Journal of Economic Perspectives* 29 (2015): 3–30.

6   C. Jara-Figueroa, A. Z. Yu, and C. A. Hidalgo, "How the Medium Shapes the Message: Printing and the Rise of the Arts and Sciences," *PLOS One* 14 (2019): e0205771.

7   E. L .Eisenstein, *The Printing Press as an Agent of Change: Communications and Cultural Trans* (Cambridge University Press, 1980).

8   Mokyr et al., "The History of Technological Anxiety."

9   *Time*, "The Automation Jobless," 24 February 1961, http://content.time.com/time /magazine/0,9263,7601610224,00.html.

10   Frey and Osborne, "The Future of Employment."

11   "The Four Global Forces Breaking All the Trends," McKinsey, https://www.mckinsey .com/business-functions/strategy-and-corporate-finance/our-insights/the-four-global -forces-breaking-all-the-trends.

12   Mokyr et al., "The History of Technological Anxiety";

Autor, "Why Are There Still So Many Jobs?";

D. Autor and A. Salomons, "Is Automation Labor Share-Displacing? Productivity Growth, Employment, and the Labor Share," *Brookings Papers on Economic Activity* (2018): 1–87;

M. Arntz, T. Gregory, and U. Zierahn, "The Risk of Automation for Jobs in OECD Countries," *OECD Social, Employment and Migration Working Papers,* No. 189 (2016), https://doi .org/10.1787/5jlz9h56dvq7-en.

13   H. J. Wilson and P. R. Daugherty, "Creating the Symbiotic AI Workforce of the Future," *MIT Sloan Management Review*, 21 October 2019, https://sloanreview.mit.edu/article /creating-the-symbiotic-ai-workforce-of-the-future/.

14   Autor, "Why Are There Still So Many Jobs?";

D. Acemoglu and P. Restrepo, "Artificial Intelligence, Automation and Work," NBER Working Paper no. 24196 (2018), http://www.nber.org/papers/w24196.

15 Autor, "Why Are There Still So Many Jobs?"

16 Autor, "Why Are There Still So Many Jobs?"

17 Frey and Osborne, "The Future of Employment."

18 Arntz et al., "The Risk of Automation for Jobs in OECD Countries."

19 Mokyr et al., "The History of Technological Anxiety";

Autor, "Why Are There Still So Many Jobs?";

Autor and Salomons, "Is Automation Labor Share-Displacing?";

Arntz et al., "The Risk of Automation for Jobs in OECD Countries";

Acemoglu and Restrepo, "Artificial Intelligence, Automation and Work."

20 Graetz and G. Michaels, "Robots at Work." *Review of Economics and Statistics* 100 (2018): 753–763.

21 L. Barbieri, C. Mussida, M. Piva, and M. Vivarelli, "Testing the Employment Impact of Automation, Robots and AI: A Survey and Some Methodological Issues," Institute for the Study of Labor (IZA) Research Paper 12612 (2019), https://papers.ssrn.com/abstract=3457656;

K. De Backer, T. DeStefano, C. Menon, and J. R. Suh, "Industrial Robotics and the Global Organisation of Production," OECD Science, Technology and Industry Working Papers 3 (2018), https://www.oecd-ilibrary.org/industry-and-services/industrial-robotics-and-the-global-organisation-of-production_dd98ff58-en.

22 Frey and Osborne, "The Future of Employment."

23 Arntz et al., "The Risk of Automation for Jobs in OECD Countries."

24 A. Agrawal, J. Gans, and A. Goldfarb, *Prediction Machines: The Simple Economics of Artificial Intelligence* (Harvard Business Review Press, 2018).

25 G. Gereffi, J. Humphrey, and T. Sturgeon, "The Governance of Global Value Chains," *Review of International Political Economy* 12 (2005): 78–104;

J. Humphrey and H. Schmitz, "How Does Insertion in Global Value Chains Affect Upgrading in Industrial Clusters?," *Regional Studies* 36 (2002): 1017–1027.

26 P.-A. Balland, C. Jara-Figueroa, S. G. Petralia, M. P. A. Steijn, D. L. Rigby, and C. A. Hidalgo, "Complex Economic Activities Concentrate in Large Cities," *Nature Human Behaviour* 4 (2020): 1–7, https://doi.org/10.1038/s41562-019-0803-3.

27 M. Roser and E. Ortiz-Ospina, "Global Education—Our World in Data," Our World in Data (2016), https://ourworldindata.org/global-education.

28 H. Rapoport, "Migration and Globalization: What's in It for Developing Countries?," *International Journal of Manpower* 37 (2016): 1209–1226.

29 R. Abbott and B. Bogenschneider, "Should Robots Pay Taxes: Tax Policy in the Age of Automation," *Harvard Law and Policy Review* (2018): 145–176.

30 A. McAfee and E. Brynjolfsson, "Human Work in the Robotic Future: Policy for the Age of Automation Essays," *Foreign Affairs* (2016): 139–150.

31 C. Jara-Figueroa, B. Jun, E. L. Glaeser, and C. A. Hidalgo, "The Role of Industry-Specific, Occupation-Specific, and Location-Specific Knowledge in the Growth and Survival of New Firms," *PNAS* 115 (2018): 12646–12653.

32 S. D. Harris and A. B. Krueger, *A Proposal for Modernizing Labor Laws for Twenty-First-Century Work: The "Independent Worker"* (Hamilton Project, Brookings, 2015).

33 A Brief History of Basic Income Ideas, https://ubi-europe.net/ubi/brief-history-basic-income-ideas/.

34 McAfee and Brynjolfsson, "Human Work in the Robotic Future."

35 A. Stern, *Raising the Floor: How a Universal Basic Income Can Renew Our Economy and Rebuild the American Dream* (PublicAffairs, 2016);

U. Colombino, "Is Unconditional Basic Income a Viable Alternative to Other Social Welfare Measures?" *IZA World of Labor* (2019), https://wol.iza.org/uploads/articles/128/pdfs/is-unconditional-basic-income-viable-alternative-to-other-social-welfare-measures.pdf;

Matt Stevens and Isabella Grullón Paz, "Andrew Yang's $1,000-a-Month Idea May Have Seemed Absurd Before. Not Now," *New York Times*, 18 March 2020, https://www.nytimes.com/2020/03/18/us/politics/universal-basic-income-andrew-yang.html

Chapter 6: Moral Functions

1 C. Efferson, R. Lalive, and E. Fehr, "The Coevolution of Cultural Groups and Ingroup Favoritism," *Science* 321 (2008): 1844–1849;

M. Hewstone, M. Rubin, and H. Willis, "Intergroup Bias," *Annual Review of Psychology* 53 (2002): 575–604;

T. Mussweiler and A. Ockenfels, "Similarity Increases Altruistic Punishment in Humans," *PNAS* 110 (2013): 19318–19323.

2 J. Haidt, S. H. Koller, and M. G. Dias, "Affect, Culture, and Morality, or Is It Wrong to Eat Your Dog?" *Journal of Personality and Social Psychology* 65 (1993): 613–628;

J. Haidt, *The Righteous Mind: Why Good People Are Divided by Politics and Religion* (Knopf Doubleday Publishing Group, 2012).

3 F. Cushman, "Crime and Punishment: Distinguishing the Roles of Causal and Intentional Analyses in Moral Judgment," *Cognition* 108 (2008): 353–380;

F. Cushman, R. Sheketoff, S. Wharton, and S. Carey, "The Development of Intent-Based Moral Judgment," *Cognition* 127 (2013): 6–21;

B. F. Malle and J. Knobe, "The Folk Concept of Intentionality," *Journal of Experimental Social Psychology* 33 (1997): 101–121;

L. Young and R. Saxe, "When Ignorance Is No Excuse: Different Roles for Intent across Moral Domains," *Cognition* 120 (2011): 202–214;

J. D. Greene, F. A. Cushman, L. E. Stewart, K. Lowenberg, L. E. Nystrom, and J. D. Cohen, "Pushing Moral Buttons: The Interaction between Personal Force and Intention in Moral Judgment," *Cognition* 111 (2009): 364–371.

Chapter 7: Liable Machines

1 I. Asimov, *I, Robot*, Robot series (Bantam Books, 1950).

2 A. Jobin, M. Ienca, and E. Vayena, "The Global Landscape of AI Ethics Guidelines," *Nature Machine Intelliegence* 1 (2019): 389–399;

V. Dignum, "Ethics in Artificial Intelligence: Introduction to the Special Issue," *Ethics and Information Technology* 20 (2018): 1–3.

3 "AI at Google: Our Principles," Google (2018), https://www.blog.google/technology/ai /ai-principles/.

4 European Commission, "Ethics Guidelines for Trustworthy AI," Digital Single Market— European Commission (2019), https://ec.europa.eu/digital-single-market/en/news /ethics-guidelines-trustworthy-ai.

5 "OECD Principles on Artificial Intelligence," Organisation for Economic Co-operation and Development (OECD), https://www.oecd.org/going-digital/ai/principles/.

6 K. Gödel, "Über formal unentscheidbare Sätze der Principia Mathematica und verwandter Systeme I," *Monatshefte für mathematik und physik* 38 (1931): 173–198;

J. Gleick, *The Information: A History, a Theory, a Flood* (Vintage, 2012).

7 P. M. Asaro, "A Body to Kick, But Still No Soul to Damn: Legal Perspectives on Robotics," in *Robot Ethics*, eds. P. Lin, K. Abney, and G. A. Bekey (MIT Press, 2011), 169–186.

8 Asaro, "A Body to Kick."

9 Asaro, "A Body to Kick."

10 Asaro, "A Body to Kick."

11 D. J. Calverley, "Android Science and Animal Rights, Does an Analogy Exist?," *Connection Science* 18 (2006): 403–417;

E. Schaerer, R. Kelley, and M. Nicolescu, "Robots as Animals: A Framework for Liability and

Responsibility in Human-Robot Interactions," in *RO-MAN 2009-The 18th IEEE International Symposium on Robot and Human Interactive Communication* (IEEE, 2009), 72–77.

12 M. Weber, *The Theory of Social and Economic Organization* (Simon and Schuster, 2009).

13 Hidalgo C. A. A Bold Idea to Replace Politicians. TED (2018), https://www.ted.com/talks/cesar_hidalgo_a_bold_idea_to_replacepoliticians?language=en.

14 S. Thrun and L Pratt, *Learning to Learn* (Springer Science & Business Media, 2012);

J. E. Tenenbaum, "Bayesian Modeling of Human Concept Learning," in *Advances in Neural Information Processing Systems,* 59–68 (1999);

J. Feldman, "The Structure of Perceptual Categories," *Journal of Mathematical Psychology 41* (1997): 145–170.

15 T. M. Mitchell, R. M. Keller, and S. T. Kedar-Cabelli, "Explanation-Based Generalization: A Unifying View," *Machine Learning* 1 (1986): 47–80.

INDEX

Page numbers in italic indicate a figure and page numbers in bold indicatea table on the corresponding page.

A

Abstract features, categorization based on, 82–83

Admissions. *See* College admissions scenarios

Agency. *See* Moral agency

Agrawal, Ajay, 109

AI. *See* Artificial intelligence (AI)

Airline reservation algorithm, 64

Alexa, privacy concerns with, 88

Algorithmic aversion, 5–6, 8, 37

Algorithmic bias, 8

   college admissions scenarios for, 70–71, 73, *75*, *79*, 83–84

   demographic correlates with, 69, 73

   equity versus predictive accuracy in, 83–85

   examples of, 64–65

   explicit versus abstract features in, 82–83

   harm-wrongness ratings of, 131

   human resource screenings scenarios for, 70, *74*, *78*

   implicit association tests of, 19–20

   Likert-type scale for, 68–69

   origins of, 64–65

   policing scenarios for, 72, *77*, *81*, 84–85

   salary increase scenarios for, 71–72, *76*, *80*

Amazon Mechanical Turk, 5n, 28, 166

Ambush scenario

   moral dimensions of, 186, *186*

   moral space representation of, 131–132

   moral space representations of, *128*, *129*, 131–132

American Airlines reservation algorithm, 64

Amusement park scenario

   moral dimensions of, 193, *193*

   moral space representation of, *128*, *129*

Anthems scenarios. *See* National symbol scenarios

Artificial intelligence (AI)

advances in, 14

interactions with, 2–3

liability for, 156–159

moral agency of, 7, 16–18, 37

moral rules versus social networks in, 150–156

moral status of, 7, 16–18

participant attitudes toward, 169–170, **171**

regulation of, 152

responsibility for actions of, 156–159

strong, 15–16

weak, 15–16

Asimov, Isaac, "three laws of robotics," 150–153

Authority

   definition of, 21

   measurement of, 28–29, **28**

   perceived levels of (*see* Moral scenarios)

Automatic associations

   cultural differences in, 22

   implicit association tests of, 19–20

Automatic teller machines (ATMs), introduction of, 107–108

Autonomous vehicle scenarios

   moral dimensions of, 23–24, 50–55, *53–54*

   moral space representation of, 127, *128*, *129*, 130, 131

Autonomous weapon systems, 14

B

Bank robbery scenario

   moral dimensions of, 185, *185*

   moral space representation of, *128*, *129*

Bias. *See also* Algorithmic bias

   intergroup, 124, 125n

   selection, 167

Bimodal judgment, 9, 146, 157

Blame, attribution of. *See also* Responsibility
  in peanut butter allergy scenario, 23
  in trolley problem, 174–176, *175*

Blasphemous comedian scenario
  moral dimensions of, 46–47, *48*
  moral space representation of, *128*, *129*

Botnik, 40–41

Brand names scenario
  moral dimensions of, 192, *192*
  moral space representation of, *128*, *129*

Brynjolfsson, E., 119

Bureaucracies, judgment of, 159–162

C

Camera bots, 88

Chatbots, 3

Citizen scoring scenario, 100–103, *101*

Civil engineer scenarios
  moral dimensions of, 201–202, *201*, *202*
  moral space representation of, *128*, *129*

Civil Rights Act (1964), 84

Cleaning robots, 17

Clopening, 106

Codelco, 50

College admissions scenarios
  efficiency versus equity in, 83–84
  moral dimensions of, 70–71, 73, *75*, *79*
  moral space representation of, *128*, *129*, 131

Comedian sketch scenario. *See* Blasphemous comedy scenario

Competence, person perception and, 82

Computer vision systems
  moral dimensions of, 91–96, *94*–*95*
  privacy concerns with, 88

Condorcet, Nicolas de, 121

Congruent trials, 19–20

Control conditions, 4

Conversational robots. *See* Chatbots

Counseling scenario
  moral dimensions of, 190, *190*
  moral space representation of, *128*, *129*

Creative artificial intelligence (AI)
  capabilities of, 40–41
  creative industry scenarios, 46–49, *48*–*49*
  limitations of, 41n
  marketing scenarios, 42–44, *44*–*45*
  moral space representation of, *128*, *129*

Culture, impact on moral judgment, 22

D

Data privacy. *See* Privacy and surveillance concerns

Deep fake videos, 40

Demographic parity, 65–66

Demographics, 9
  algorithmic bias and, 69, 73
  effects on moral judgment, 141–146, **142–143**, *145*
  of study participants, 167–169, **168**

Desecration of national symbols. *See* National symbol scenarios

Detroit (game), 153–154

Differential privacy, 90–91

Digital records, privacy concerns with, 88

Discriminatory situations. *See also* Algorithmic bias
  college admissions, 70–71, 73, *75*, *79*, 83–84
  demographic correlates with, 69, 73
  fairness in, 64–68
  human resource screenings, 70, *74*, *78*
  Likert-type scale for, 68–69
  moral space representation of, *128*, *129*
  policing, 72, *77*, *81*
  salary increases, 71–72, *76*, *80*

Displacement. *See* Labor displacement

Domingos, Pedro, 17

Drones, military use of, 14

Durkheim, Emile, 160

E

Education levels
    effects on moral judgment, 9, 141–146, **142–143**, *145*
    of study participants, 167–169, **168**
Emotions, in moral reasoning, 19–22
Enlightenment, harm basis of morality in, 18–19
Equality in false acceptances, 66
Equality of false rejections, 65–66
Ethical Algorithm, The (Kearns and Roth), 90
Ethical concerns, 10, 16–18. *See also* Algorithmic bias; Labor displacement; Privacy and surveillance concerns
    autonomous vehicles, 23–24, 50–55, *53–54*
    lewd or disrespectful behavior, 40–49, 41n, *44–45*, *48–49*
    moral agency, 7, 16–18, 37
    moral status, 7
    national symbol desecration, 56–61, *58–59*
    need for research into, 3–6
    normative versus positive approaches to, 5–6
    risk-taking behavior, 35–38, *38–39*, 178–179, *178–179*, 180–181, *180–181*
    strong versus weak AI in, 15–16
Ethnicity
    correlation with moral judgments, 141–146, **142–143**, *145*
    of study participants, 167–169, **168**
European Commission, 152
Explicit features, categorization based on, 82–83
Explosive Ordinance Disposal (EOD) robots, 17

F

Facial recognition systems
    algorithmic bias in, 67

moral dimensions of, 91–96, *94–95*
Facial recognition technology, 14
Fairness, 21. *See also* Algorithmic bias
    equality of false rejections, 65–66
    measurement of, 28–29, **28**
    perceived levels of (*see* Moral scenarios)
    statistical parity, 65–66
Fake news, 40
False acceptances, equality in, 66
False rejections, equality of, 65–66
Federated Learning, 91
First Amendment rights, 56
"Flag-burning" amendment, 56
Flag desecration scenarios
    moral dimensions of, 56–57, *58–59*
    moral space representation of, *128*, *129*
Foreign contractors, displacement attributed to, 113–118, *117*
Foreign subsidiaries, displacement attributed to, 113–118, *117*
Foreign temporary workers, displacement attributed to, 109–118, *112–113*, *117*
Forest fire scenario
    moral dimensions of, 178–179, *178–179*
    moral space representation of, *128*, *129*, 130
Frankenstein (Shelley), 2, 162
Functions, moral. *See* Moral functions

G

Gaby mine, 50
GANs. *See* Generative Adversarial Networks (GANs)
Gas tax scenario
    moral dimensions of, 197, *197*
    moral space representation of, *128*, *129*
Gender
    correlation with moral judgments, 9, 141–146, **142–143**, *145*
    of study participants, 167–169, **168**

Generative Adversarial Networks (GANs), 40
Generative AIs, 40–41
Gödel, Kurt Friedrich, 153
Good prediction scenario. *See* Procurement management scenarios
Google, 152
Graetz, G., 108
Graveyard excavation scenario
  moral dimensions of, 26–31, *29*
  moral space representation of, 127, *128, 129*
Griggs v. Duke Power Company (1971), 84

H

Haidt, Jonathan, 125
Harm, 21. *See also* Intentionality
  measurement of, 28–29, **28**
  moral functions of, 133–141, *137, 138, 139, 140*
  moral space representation of, 125–132, *126, 128, 129*
  patterns across scenarios, 127–132
  perceived levels of (*see* Moral scenarios)
Harm basis of morality, 18–19
Harm-intention plane, 127, *128, 129*, 130
Harm-wrongness plane, *129*, 131–132
Harris, Seth, 119
Heroic assumptions, 133
Heuristics, biases as, 82
Hikers scenario
  moral dimensions of, 191, *191*
  moral space representation of, *128, 129*
Hiring decisions. *See* Human resource screenings scenarios
HIS Group hotel chain, hacking of, 88
Human behavior, judgment of. *See* Moral judgments
Human learning, limitations of, 161–162
Human resource screenings scenarios
  moral dimensions of, 70, *74, 78*

moral space representation of, *128, 129,131*
Hume, David, 19n
Hurricane scenario
  moral dimensions of, 180–181, *180–181*
  moral space representation of, *128, 129*, 130

I

I, Robot (Asimov), 150–152
Implicit association tests, 19–20
Implicit.harvard.edu website, 20
Incongruent trials, 19–20
Independent worker classification, 119
Industrial Revolution, 107
In-group favoritism. *See* Intergroup bias
Intentionality, 10, 159–161
  demographic correlates with, 141–146, **142–143**, *145*
  human behavior judged by, 10, 23–26, *25*, 124, 139, 146–147, 157, 159–162
  moral functions of, 133–141, *137, 138, 139, 140*
  moral space representation of, 125–132, *126, 128, 129*
  patterns across scenarios, 127–132
  perceived levels of (*see* Moral scenarios)
  wrongness and, 23–26, *25*
Interest rate scenario
  moral dimensions of, 198, *198*
  moral space representation of, *128, 129*, 131
Intergroup bias, 124, 125n
International Federation of Robotics (IFR), 108
Intuition
  cultural differences in, 22
  implicit association tests of, 19–20

J

Jewelry robbery scenario
  moral dimensions of, 184, *184*

moral space representation of, *128, 129*
Judgments. *See* Moral judgments

K

K-anonymity, 89–90
Kearns, Michael, 90
Kleinberg, Jon, 83
Knowledge diffusion, labor mobility as channel of, 119
Komatsu, 50
Kronos, 106
Krueger, Alan, 119

L

Labor displacement
    factors affecting, 109
    fear of, 107–108
    inequality and, 14
    interaction between technology and labor in, 107–109
    mitigation of consequences of, 118–121
    moral dimensions of, 9, 109–118, *112–113, 117*, 156
    precarization of work and, 106–107
Labor mobility, barriers to, 119
Learning, human versus machine, 161–162
Legal implications, 10, 156–159
Lewd behavior
    creative marketing scenarios, 42–44, *44–45*
    moral space representation of, *128, 129*
    playwright scenario, 47, *49*
Liability for machine actions, 156–159
Licensing, state, 119
Life-or-death decisions, attitudes toward AI in
    forest fire scenario, *178–179*, 178–179
    hurricane scenario, 180–181, *180–181*
    moral space representation of, *128, 129*
    tsunami scenario, 35–38, *38–39*

Likability
    demographic correlates with, 141–146, **142–143**, *145*
    perceived levels of (*see* Moral scenarios)
Likert-type scales
    for algorithmic bias scenarios, 68–69
    demographic correlates with, 141–146, **142–143**, *145*
    for graveyard excavator scenario, 26–27
    for labor displacement scenarios, 111
    for privacy scenarios, 94
Literai, 40
Looms, Luddite opposition to, 4, 107
Loyalty
    definition of, 21
    measurement of, 28–29, **28**
    perceived levels of (*see* Moral scenarios)
Luddites, opposition to steam-powered looms, 4, 107
Ludwig, Jens, 83

M

Machine actions, human judgment of. *See* Moral judgments
Machine learning, 161–162
Malle, B. F., trolley problem of, 4–5, 4n
    blame attributions for, 174–176, *175*
    replication of, 172–173
    wrongness attributions for, 173–174, *174*
Marketing scenarios
    moral dimensions of, 42–44, *44–45*
    moral space representation of, *128, 129*
Marx, Karl, 160
Mathematics, incompleteness of, 153
McAfee, A., 119
Mechanical Turk (MTurk), 5n, 28, 166
Methodology. *See* Study methodology
Michaels, G., 108

Microsoft Tay bot, 3
Military draft, moral dimensions of, 21–22
MIT Committee on the Use of Humans as Experimental Subjects, 166
Mobile phone traces, 89
Moral agency, 7, 16–18, 37
Moral dimensions
    components of, 20–22
    impact on moral judgment, 22–26, *25*
    intentionality and, 23–26
    measurement of, 28–29
    perceived levels of (*see* Moral scenarios)
Moral foundation theory, 125
Moral functions. *See also* Moral space
    definition of, 124–125
    modeling of, 133–141, **134**, **135**, *137*, *138*, *139*, *140*
Morality. *See also* Moral judgments; Moral scenarios
    cultural differences in, 22
    definition of, 18
    dimensions of, 20–29, *25*
    emotions and automatic associations in, 19–22
    Enlightenment theory of, 18–19
    harm basis of, 18–19
    implicit association tests of, 19–20
    moral foundation theory of, 125
Moral judgments, 18–19. *See also* Moral scenarios
    algorithmic aversion and, 5–6, 8, 37
    bimodal, 9, 146, 157
    demographic correlates of, 9, 141–146, **142–143**, *145*
    ethical and legal implications of, 10, 156–159
    intentionality in, *25*
    intentions versus outcomes in, 10, 23–26, *25*, 124 139, 146–147, 157, 159–162 (*see also* Intentionality)
    mathematical representation of, 24–26, *25*
    moral functions of, 124–125, 133–141, **134**, **135**, *137*, *138*, *139*, *140*
    moral space representation of, 125–132, *126*, *128*, *129*
    normative approach to, 5
    positive approach to, 5–6
    rejection/acceptance of technology and, 6
    responsibility for machine actions in, 156–159
    statistical models of, 9–10
    trolley problem example, 4–5, 4n, 172–173
    unimodal, 9, 146, 157
Moral rules, contradictions with social relationships, 150–156
Moral scenarios
    ambush, 128, 129, 131–132, 186, *186*
    amusement park, *128*, *129*, 193, *193*
    autonomous vehicles, 23–24, 50–55, *53–54*, 127, *128*, *129*, 130–131
    bank robbery, *128*, *129*, 185, *185*
    blasphemous comedian, 46–47, *48*, *128*, *129*
    brand names, *128*, *129*, 192, *192*
    citizen scoring, 100–103, *101*
    civil engineer, *128*, *129*, 201–202, *201*, *202*
    college admissions, 70–71, 73, *75*, *79*, 83–84, *128*, *129*, 131
    computer vision systems, 91–96, *94–95*
    counseling, *128*, *129*, 190, *190*
    creative marketing, 42–44, *44–45*, *128*, *129*
    definition of, 26
    dilemmas versus, 26
    forest fire, *128*, *129*, 130, 178–179, *178–179*
    gas tax, *128*, *129*, 197, *197*
    graveyard excavation, 26–31, **28**, *29*, *30*, 127, *128*, *129*
    hikers, *128*, *129*, 191, *191*
    human resource screenings, 70, *74*, *78*, *128*, *129*, 131
    hurricane, *128*, *129*, 130, 180–181, *180–181*
    interest rate, *128*, *129*, 131, 198, *198*
    jewelry robbery, *128*, *129*, 184, *184*
    labor displacement, 9, 109–118, *112–113*, *117*, 156
    lewd playwright, 47, *49*, *128*, *129*

mathematical representation of, 24–26, *25*

moral agency in, 7, 16–18, 37

national symbol desecration, 56–61, *58–59*, 127, *128*, *129*

nursing assistant, *128*, *129*, 194, *194*

patterns across, 127–132

peanut butter allergy, 23–26, *25*

personal assistant, *128*, *129*, 200, *200*

pharmacy robbery, *128*, *129*, 183, *183*

physical therapist, *128*, *129*, 195, *195*

plagiarizing songwriter, 46–47, *48*, 128, *129*

policing, 72, *77*, *81*, 84–85, *128*, *129*

procurement management, *128*, *129*, 187–189, *187*, *188*, *189*

recommender systems, 96–100, *98–99*

salary increases, 71–72, *76*, *80*, *128*, *129*

shoplifting security, *128*, *129*, 203, *203*

supermarket robbery, *128*, *129*, 182, *182*

suspected terrorist, *128*, *129*, 131–132, 199, *199*

tsunami, 35–38, *38–39*, *128*, *129*, 130–131

Twitter, *128, 129*, 196, *196*

Moral space

definition of, 125–126

example of, 126–127, *126*

harm-intention plane in, 127, *128*, *129*, 130

harm-wrongness plane in, *129*, 131–132

wrongness-intention plane in, *129*, 130–131

Moral status, 7, 16–18

Moral surface, 133–141

Moral wrongness. *See* Wrongness, perceived levels of

More, Thomas, 121

Mullainathan, Sendhil, 83

N

National symbol scenarios

moral dimensions of, 56–61, *58–59*

moral space representation of, 127, *128*, *129*

Navarro, Jannette, 106

Negligence, 158

Normative approaches, 5–6

Nursing assistant scenario, 194, *194*

O

OECD. *See* Organisation for Economic Co-operation and Development (OECD)

Offshoring, displacement attributed to, 113–118, *117*

Online dating systems, 96–100, *99*

Organisation for Economic Co-operation and Development (OECD), 152

Outcomes, moral judgment based on, 10, 23–26, *25*, 124, 139, 146–147, 157, 159–162

Outsourcing, displacement attributed to, 113–118, *117*

P

Parity, statistical, 65–66

Participants

attitudes toward AI, 169–170, **171**

demographic characteristics of, 167–169, **168**

recruitment of, 28, 166–167

PATE, 91

Peanut butter allergy scenario, 23–26, *25*

Personal assistant scenario

moral dimensions of, 200, *200*

moral space representation of, 128, *129*

Pharmacy robbery scenario

moral dimensions of, 183, *183*

moral space representation of, *128*, *129*

Physical therapist scenario

moral dimensions of, 195, *195*

moral space representation of, *128*, *129*

Plagiarizing songwriter scenario

moral dimensions of, 46–47, *48*

moral space representation of, *128, 129*
Policing scenarios
   moral dimensions of, 72, *77, 81*
   moral space representation of, *128, 129*
   outcome-based approach to, 84–85
Political orientation of study participants, 168–169
Positive approaches, 5–6
Precarization of work, 106–107
Prediction costs, 109
Prediction Machines (Agrawal), 109
Predictive purchasing systems, 96–100, *98*
Prejudice, 32
Pretrial risk assessment tools, algorithmic bias in, 67–68
Principal components, 82–83
Printing, introduction of, 4, 107
Privacy and surveillance concerns, 14, 156
   citizen scoring scenario, 100–103, *101*
   computer vision systems scenarios, 91–96, *94–95*
   Likert-type scale for, 94
   privacy-preserving algorithms, 91
   proposed solutions to, 89–91
   recommender system scenarios, 96–100, *98–99*
   risks of AI to, 64–65
Procurement management scenarios
   moral dimensions of, 187–189, *187, 188, 189*
   moral space representation of, *128, 129*
Product liability, 158–159
Promotion decisions. *See* Salary increase scenarios
Purity
   definition of, 21
   measurement of, 28–29, **28**
   perceived levels of (*see* Moral scenarios)
P-values, 30–31
   ambush scenario, *186*
   amusement park scenario, *193*
   autonomous vehicle scenarios, *53–54*
   bank robbery scenario, *185*
   blasphemous comedian scenario, *48*

   brand names scenario, *192*
   citizen scoring scenario, *101*
   civil engineer scenarios, *201, 202*
   college admissions scenarios, *75, 79*
   computer vision systems scenarios, *94–95*
   counseling scenario, *190*
   creative marketing, *44–45*
   forest fire scenario, *178–179*
   gas tax scenario, *197*
   graveyard excavation scenario, *30*
   hikers scenario, *191*
   human resource screenings scenarios, *74, 78*
   hurricane scenario, *180–181*
   interest rate scenario, *198*
   jewelry robbery scenario, *184*
   labor displacement scenarios, *112–113, 117*
   lewd playwright scenario, *49*
   national symbol desecration, *58–59*
   nursing assistant scenario, *194*
   personal assistant scenario, *200*
   pharmacy robbery scenario, *183*
   physical therapist scenario, *195*
   plagiarizing songwriter scenario, *48*
   policing scenarios, *77, 81*
   procurement management scenarios, *187, 188, 189*
   recommender systems scenarios, *98–99*
   salary increase scenarios, *76, 80*
   shoplifting security scenario, *203*
   supermarket robbery scenario, *182*
   suspected terrorist scenario, *199*
   tsunami scenario, *38–39*
   Twitter scenario, *196*

Q

QR codes, 108

R

Rambachan, Ashesh, 83

Randomized response, 90–91

Rappor, 91

Rationality, morality and, 18–19

Raw material shortage/excess. *See* Procurement management scenarios

Real-world studies, 7

Recidivism, 67

Recklessness, 158

Recommender systems scenarios, 96–100, *98–99*

Regulation, labor, 118–119

Reidentification risks, 89

Religious orientation

    effects on moral judgment, 141–146, **142–143**, *145*

    of study participants, 167–169, *168*

Replace different/same assessments

    demographic correlates with, 141–146, **142–143**, *145*

    of scenarios (*see* Moral scenarios)

Research methodology. *See* Study methodology

Responsibility

    liability and, 156–159

    perceived levels of (*see* Moral scenarios)

    trolley problem example, 174–176, *175*

Rio Tinto, 50

Risk-taking, attitudes toward AI in

    forest fire scenario, 178–179, *178–179*

    hurricane scenario, 180–181, *180–181*

    moral space representation of, 130–131

    tsunami scenario, 35–38, *38–39*

Robbery scenarios

    moral dimensions of, 182–185, *182*, *183*, *184*

    moral space representation of, *128*, *129*

Robots

    camera bots, 88

    chatbots, 3

    Explosive Ordinance Disposal, 17

    health-care, 14

    moral agency of, 7, 16–18, 37

    moral status of, 7, 16–18

    tax on use of, 119

Roth, Aaron, 90

S

Salary increase scenarios

    moral dimensions of, 71–72, *76*, *80*

    moral space representation of, *128*, *129*

Scenarios. *See* Moral scenarios

Selection bias, 167

Self-driving cars. *See* Autonomous vehicle scenarios

Sex robots, 17

Shelley, Mary, 2, 162

Shelley AI, 40–41

Shoplifting security scenario

    moral dimensions of, 203, *203*

    moral space representation of, *128*, *129*

Similar situation scale

    demographic correlates with, 141–146, **142–143**, *145*

    perceived levels of (*see* Moral scenarios)

Simulated studies, 7

Siri, privacy concerns with, 88

Social networks, contradictions arising from, 150–156

Split Learning, 91

Starbucks, 106

State licensing, 119

Statistical models, 9–10

Statistical parity, 65–66

Status, moral, 7, 16–18

Stereotyping, 82

Strong artificial intelligence, 15–16

Study methodology, 3–4, 8, 26–31. *See also* Trolley problem

Likert-type scales, 26–27, 68–69, 94, 111
participant attitudes, 169–170, **171**
participant demographics, 167–169, **168**
participant recruitment, 28, 166–167
scenario structure, 26
study design, 166
word association exercise, 28–29, **28**
Supermarket robbery scenario
moral dimensions of, 182, *182*
moral space representation of, *128, 129*
Surveillance. *See* Privacy and surveillance concerns
Suspected terrorist scenario
moral dimensions of, 199, *199*
moral space representation of, *128, 129*, 131–132
Sweeney, Latanya, 89

T

Tax, robot, 119
Tay chatbot (Microsoft), 3
Technological displacement. *See* Labor displacement
Temporary foreign workers, displacement attributed to, 109–118, *112–113, 117*
Texas v. Johnson (1989), 56
"Three laws of robotics" (Asimov), 152
Travel company scenario, 96–100, *99*
Trolley problem
blame attributions for, 4–5, 174–176, *175*
description of, 4n, 172–173
moral status in, 17
wrongness attributions for, 173–174, *174*
Tsunami scenario
moral dimensions of, 35–38, *38–39*
moral space representation of, *128, 129*, 130–131
Tweeting bots, 40
Twitter scenario
moral dimensions of, 196, *196*
moral space representation of, *128, 129*

U

Unfair treatment. *See* Discriminatory situations
Unimodal distribution, 9, 146, 157
United Airlines reservation algorithm, 64
Universal basic income (UBI), 121
Unlearning, 161–162

V

Vicarious liability, 158–159

W

Warmth, person perception and, 82
Weak artificial intelligence (AI), 15–16
Weber, Max, 159, 160
Weld, William, 89
Word association exercise, 28–29, **28**
Word embedding, 67
Wrongness, perceived levels of. *See also* Intentionality; Moral scenarios
blame attribution, 23, 174–176, *175*
demographic correlates with, 141–146, **142–143**, *145*
moral functions of, 133–141, *137, 138, 139, 140*
moral space representation of, 125–132, *126, 128, 129*
patterns across scenarios, 127–132
in trolley problem, 173–174, *174*
Wrongness-intention plane, *129*, 130–131

Y

Yang, Andrew, 121
Younger workers, displacement attributed to, 113–118, *117*